THE VOICE CATCHERS

THE VOICE CATCHERS

HOW MARKETERS LISTEN IN
TO EXPLOIT YOUR FEELINGS,
YOUR PRIVACY, AND YOUR WALLET

JOSEPH TUROW

Yale UNIVERSITY PRESS

NEW HAVEN & LONDON

Published with assistance from the foundation established in memory of
James Wesley Cooper of the Class of 1865, Yale College.

Yale University Press books may be purchased in quantity for educational,
business, or promotional use. For information, please e-mail sales.press@
yale.edu (U.S. office) or sales@yaleup.co.uk (U.K. office).

Set in Meridien and Futura types by IDS Infotech Ltd.
Printed in the United States of America.

Library of Congress Control Number: 2020947840
ISBN 978-0-300-24803-6 (hardcover : alk. paper)

A catalogue record for this book is available from the British Library.

This paper meets the requirements of ANSI/NISO Z39.48–1992
(Permanence of Paper).

10 9 8 7 6 5 4 3 2 1

For Judy

CONTENTS

THE VOICE CATCHERS

INTRODUCTION

Here Comes the Voice Intelligence Industry

Your voice is unique. No one else has it. And because your voice belongs to no one else, it's extraordinarily valuable, not only to you, but also to a new sector of society that is designed to exploit it: the voice intelligence industry. Built by marketers to collect information from the ways individuals talk and sound, this industry is deploying immense resources and breakthrough technologies based on the idea that voice is biometric—a part of your body that those in the industry believe can be used to identify and evaluate you instantly and permanently. Companies are working to analyze your vocal-cord sounds and speech patterns for information about your emotions, sentiments, and personality characteristics, all so that they can better persuade you, often in real time. Soon they may be able to draw conclusions about your weight, height, age, ethnicity, and more—all characteristics that scientists believe are revealed by your voice. Marketers will be able to score you as more or less valuable, show you different products based on that valuation, give you discounts that are better or worse than the ones they give other people, and treat you better or worse than others when you want help. In other words, marketers are using voice data to model ways to discriminate between you and others in unprecedently powerful ways. And all of this is happening without adequate regulations and safeguards to help American consumers understand the risks. The aim of this book is to describe this developing domain, explain how it's already influencing our lives, and show what about it needs to be stopped. Now is the time to promote perspectives and policies to derail the voice-based world of marketing biometrics—while the industry is still being built, and before socially corrosive processes linked to it become too entrenched to change.

The emerging voice intelligence industry involves such tools as smart speakers, car information systems, customer service calls, and "connected-home" devices like thermostats and alarms. When you talk, their "intelligent assistants" can draw inferences about you using analytical formulas generated by artificial intelligence. In the United States and European Union, the best-known assistants tasked with performing such activities are Amazon's Alexa, Google Assistant, and Apple's Siri. In China, Baidu is doing it with its DuerOS voice assistant, and Alibaba with Tmall Genie. Each carries out its work through tens of millions of smart speakers (WiFi linked audio devices), smartphones, and car audio systems.[1] Google Assistant, accessed mostly through smartphones and Google Home products, is now available in more than a billion devices.[2] Amazon claims that its Alexa personal assistant is present in "hundreds of millions" of devices.[3] A different, though related, ecosystem of firms is creating voice initiatives propelled by artificial intelligence in customer contact centers.

Public attention to the voice industry has centered primarily on smart speakers. Dubbed "voice first" devices by marketers, these are cylinders (or more recently other shapes) that sometimes come with screens. Ask a question or make a request, and the devices can access a huge number of information sources including app-like add-ons contributed by thousands of companies, nonprofits, and even individuals. Owners most typically use the devices to check the weather, set timers, learn recipes, listen to music, play games, ask for facts, and buy things. In the United States the explosion of smart-speaker sales began around 2014 with the introduction of Amazon's Echo and its assistant, Alexa. The Google Home came out almost exactly two years later, and then smart speakers from other firms came tumbling out. Apple and Samsung used preexisting assistants (Siri for Apple, Bixby for Samsung), and companies like Sonos built speakers that link to Alexa or Google Assistant, or both. Press attention during this period has see-sawed between the latest capabilities built into these devices and the new social dangers they represent. Many stories center on the smart speaker's ability to "listen" and then answer. The gizmo starts recording whenever it hears the wake word ("Alexa," "Hey Google," "Siri"), and it tracks sound for up to sixty seconds each time. Ask "Alexa, what's the temperature in Chicago?" and a (so far) immutable woman's voice will provide a direct response. Try "Hey Google, how many plays did Shakespeare write?" and a female voice (which in this case you can change to male) will give a concise (and correct) answer (thirty-seven), along with two sentences that elaborate.

But there have been incidents. A six-year-old girl in Texas used Alexa to order a $170 dollhouse and four pounds of sugar cookies—by simply asking for them.[4] An Amazon user in Germany requested data about what he had said to the device, and he instead received 1,700 audio recordings of someone he didn't know.[5] Alexa users have reported bursts of unexpected and scary laughter coming from the cylinder.[6] A woman in Portland, Oregon, found out that the Echo had recorded a conversation she had had with her husband without the couple's knowledge, then sent the recording to a random person on their contacts list.[7] Amazon had rational explanations for all these issues—a parent not setting the purchase-protection code, an Amazon employee's error, Alexa's mistakenly hearing a command to laugh, a rare combination of speaking-and-listening accidents. "As unlikely as this string of events is, we are evaluating options to make this case even less likely" was the official comment on the Portland couple's case, but it could have been the same for any of them. The voice firms are also playing whack-a-mole with hackers and trying to rid the system of bugs that could open the data to snoopers. A writer wryly described one of the privacy-intrusion incidents as "the latest nightmare scenario for the tech-phobic."[8]

The real difficulties with the smart speakers and the voice intelligence industry, however, have yet to emerge. The unwanted incidents will come not from bugs, hacks, or glitches, but from features of technology that work properly. That's because the system is evolving into a blueprint for marketers to use your body's signals for gain. Consider:

- A cartoon drawing accompanying an Amazon patent depicts a woman "coughing" with a "sniffle" as she tells one of Amazon's smart speakers, "I'm hungry." The device picks up speech irregularities that imply a cold ("based at least in part on an analysis of pitch, pulse, voicing, jittering, and/or harmonicity of a user's voice, as determined from processing the voice data"). Based on that conclusion, Alexa asks if the person wants chicken soup, and when she says no, offers to sell her cough drops with one-hour delivery.[9] The scenario may sound helpful, but learning how often someone will need to drink chicken soup and agree to buy cough drops can lead to an AI program drawing inferences about a person's short- or long-term health. These conclusions would have marketing value. Knowing via voice if someone is sick could benefit Amazon Pharmacy, the firm's advising, ordering, and delivery service for prescription medicines.[10]

- Another patent has Alexa listening through a smart speaker for "keywords" such as *enjoyed* or *love*. When it hears a trigger word, it "captures adjacent audio that can be analyzed on the device or remotely," to figure out what the person enjoyed or loves; the individual might say *I enjoy traveling to San Francisco* or *I love hip-hop* or *I love Judy*. Tracking the keywords would allow Amazon to add information to people's profiles so it can sell them items related to what they like and not what they dislike, and sell advertisers the ability to reach people with messages that reflect those sentiments. Amazon and its advertisers may also avoid making offers to people who say they love or enjoy what the advertisers disapprove of, or who for personal or cultural reasons don't use those specific words to express happiness.[11]

- A Google patent application describes the firm's ability to use "characteristics of audio signatures, such as speech patterns, pitch, etc.," to figure out who is in a room, whether they are "moving or performing other actions," and how quietly they are doing it. Google describes a situation where parents can know from afar whether their children are sleeping or whispering. The latter, says Google, would indicate "mischief . . . is occurring," and the system would notify parents and other adults so they could "exercise control." The patent clearly aims at building the firm's "smart home" business, an enterprise centering around devices like lamps, thermostats, and locks that respond to an owner's commands through voice and touch.[12]

Two Amazon representatives who wanted anonymity told me it is company policy not to comment about patents. Both also said patents take a long time to bear fruit. That should not prevent our discussing them. Amazon, Google, and other voice intelligence firms are in business for the long term, and our society will likely continue to be influenced by their innovations for generations to come. In fact, as if to underscore the utility of those patents, Amazon announced during fall 2020 that its just-released Halo health and wellness band is able to analyze the tone of its owner's voice for "qualities . . . like energy and positivity."[13] Amazon declared that getting people to consider the emotions that their voice emits will encourage them to adopt healthier communication practices with their loved ones and bosses. The company asserted that the Halo's security features would make its analysis off-limits to anyone but the person speaking; the voice profile, too, is explicitly not for use by third parties. Yet in the face of all the developments you'll see in this book, it is hard not to understand the Halo's professed capability as a proof of concept. The entire voice profiling idea demonstrated here can, as the patents suggest, easily be ported to the marketing realm and beyond.

Moreover, the building blocks for the patent scenarios are already in place. Voice discrimination already goes beyond what Amazon and Google do. The customer phone service (or "contact center") business was first out of the gate in profiting from individuals' unique voices. Contact center firms such as Nuance and Verint already evaluate a caller's sounds and linguistic patterns for emotion, sentiment, and personality. Linking those biometrics with the caller's name, the firms regularly tell reps to give discounts to tense-sounding customers who are big spenders in order to mollify them. Contact-center software also routinely shunts customers pegged as "talkative" to reps with a track record of getting along with such people and of getting them to spend extra money ("upselling" them).[14]

The voice-intelligence industry argues that the assistants are programmed only to help people in a range of everyday activities on the phone, at home, in cars, in stores, and on streets. But they are tied to advanced machine learning and deep neural-network programs that companies can use to analyze what individuals say and how they say it with the goals of discovering when and whether particular people are worth persuading, and then finding ways to persuade those who make the cut. Amazon and Google, the highest-profile forces in voice today, are not yet using the maximum marketing potential of these tools, evidently because they are worried about inflaming social fears around the collection of people's voices. But contact centers, which are out of the public eye and thus more audacious about dealing differently with people based on how they talk, may represent the future. As people get used to giving up their voice virtually everywhere, stores, banks, and other sellers that develop their own voice assistants will have fewer qualms about exploiting what their customers say and how they say it. Google and Amazon may ultimately feel compelled to join them.

We accept marketers' assurances about voice-information privacy at our peril.

Marketers have approached voice as an exploitable part of the human body that doesn't have the negative associations of facial recognition. This relatively blank slate, they believe, offers an opportunity to cultivate trust. Yet the voice profiling activities that already happen every day raise many ethical issues. Should it be socially acceptable when what people say to their voice assistants and how they say it is being preserved by the assistant's firm for purposes we don't know—and that the firm might not know either? Is it okay when a call-center computer senses a caller's anger and posts real-time messages to the service representatives about what words will likely calm the caller? Or when, as is the aim of a U.S. patent application, a bank's call center analyzes the emotions in a person's voice to decide whether that person is a good risk for a loan?[15]

Also worrisome are unanswered questions related to political power. The prospect of tracking people and controlling them is as appealing to political campaigns and government agencies as it is to commercial marketers. Facial recognition surveillance—a stock-in-trade of governments for public security—doesn't yet reach into the recesses of people's homes and personal lives, but voice does. It shouldn't be surprising, then, that as voice profiling becomes the first technology system to reach into our public and private lives with biometric analysis, political marketers and governments are taking notice. Consider just a few possibilities:

- A political campaign that can sense people's voice responses to the political message at the beginning of a phone call and use that data to decide on the fly how to describe that candidate's beliefs.

- A political marketer who streams certain commercials only to those individuals whose comments and voice characteristics suggest an openness to extreme arguments the marketer wouldn't want others to see.

- A police department that arrests someone because publicly placed microphones note a voice pattern predictive of subsequent violence when he mentions an elected official in conversation.

- A government agency that decides whom to imprison—and torture—not only because of what the accused said, but how he or she said it.

Such consequences may seem hard to accept in these early days of the voice intelligence era, when assistants like Amazon's Alexa, Google Assistant, Apple's Siri, Microsoft's Cortana, Samsung's Bixby, Nuance's Rita, and Bank of America's Erica seem so engaging, helpful, and harmless. The personal and social costs of these technologies will emerge, however. The voice intelligence industry is in what several insiders I interviewed for this book called a scale-building period. Each brand of intelligent agent is working to create technologies and experiences that will seduce huge numbers of people to become habituated to its offerings. ("We basically envision a world where Alexa is everywhere," one Amazon executive put it in 2018.)[16] Once the company becomes a firm part of many people's lives, and once its engineers have figured out the best ways to turn what it knows about its customers into profits, the scale-building period will end, and the hardcore use of their voice data for profiling and discriminatory offers will begin. Hewing to the letter (if not the spirit) of whatever privacy laws exist, the company will roar ahead in its new incarnation, even if most of its users joined when that business model did not yet exist. This classic bait-and-switch marked the rise of both Google and Facebook. Both companies attracted colossal numbers of adherents when their business models did not require manipulating massive numbers of data points about them. Facebook especially benefited from the network effect—friends' participation means that you're likely to use that social network more than others. Only when the numbers flocking to these sites became large enough to attract high-paying advertisers did their business models solidify around selling ads personalized to what Google and Facebook knew about their visitors. By then, the sites had become such important parts of users' daily activities that they felt they couldn't leave, despite their concern about data flows they didn't understand and couldn't control.

If we look carefully at where voice firms today are building scale and where other marketers are pushing to piggyback on that scale, we can begin to understand future business uses of voice, as well as what political campaigns, police, and government agencies could do with these tools. In later chapters, with these goals in mind, I will unpack attempts by the major voice companies and marketers to install, implement, celebrate, and profit from voice intelligence technologies. First, however, I will use a social framework to help explain why marketers are chasing voice, and what an examination of the voice intelligence industry can tell us about surveillance generally.

In recent years there has been a great deal of scholarly and journalistic investigation of social monitoring and privacy.[17] Much has been written, too, about the consequences of surreptitious tracking and unchecked data collection, plus the non-transparent, often discriminatory use of information acquired without knowledge or consent.[18] But the forces that lead audiences to be ensnared in these activities have received less attention. We know little about how industries in key sectors of public life seduce populations into using technologies for the industries' surveillance aims, how they turn digital surveillance into habits of everyday life—and then make uncoupling from them extremely difficult. In investigating these processes in this book, I will be exploring an industry whose creation is still playing out. By calling attention to these developments while the voice surveillance industry is still in a formative stage, I hope to make it easier for activists and citizens to stop at least the most objectionable practices before they become too ingrained in our everyday lives.

Throughout, I will keep an eye on four related corporate strategies that are shaping the evolving industry. One is the *spiral of personalization* that drives much of twenty-first-century marketing. Another, *seductive surveillance*, describes how companies attract people to their technologies by playing up the devices' allure and playing down their creepy features. The third, *habituation*, explores how companies help people become accustomed to their tracking technologies. And the fourth, *resignation*, refers to how the industry works to make sure that, despite their worries, people continue to use these technologies because they believe digital surveillance is unavoidable. Taken together, the analysis of these strategies can help us, the users of these technologies, understand what is going on and what we can do about it.

The spiral of personalization is the guiding spirit behind the new marketing order. Marketers believe that in order to remain competitive, they must gather as much data as possible about current and prospective customers and send them individualized messages and product offerings.[19] Yet because of the very nature of this work, marketers never feel they are doing enough to know their targets and reach them efficiently. They will always be looking ahead to technologies like voice intelligence that promise new forms of information made possible by greater intrusions into people's lives. That is, today's new technology will soon also fail to satisfy the drive to uncover deeper knowledge of the customer, leading to the next innovation—for instance, linking voice intelligence to even more intrusive technologies. This unending spiral of personalization will, if left unchecked, ratchet personal surveillance higher and higher with little attention to the social costs, chief of which are the continued erosion of anonymity as a value and a decline in the very freedom of choice that the industry contends is its greatest gift to its customers.[20]

As a marketing approach, personalization is both old and new. Street peddlers have always used their understanding of their customers to sell goods. One way to do that was to cover a limited territory and return regularly to the same villages, where they could develop ongoing relationships with individual customers. The peddlers offered their goods to buyers they knew, and after haggling to reach an acceptable price, they typically took notes about the final deal so they would be consistent from one sale to the next. The repeated transactions motivated buyers and sellers alike toward honesty and fair dealing. Starting in the nineteenth century, the peddling business in North America slowed as shoppers turned to small dry-goods emporia, grocery stores, and "general" stores, but the selective presentation and pricing of goods based on the proprietor's knowledge of buyers continued. By the late nineteenth and early twentieth centuries, though, these small operations increasingly gave way to department stores and self-service food markets with many clerks whom the owners didn't necessarily trust to judge people well or bargain successfully. To get around that difficulty, proprietors put out all the goods they sold for everyone to see, and they posted prices that everyone would have to pay. The result was a twentieth-century "democratized" retail environment: anyone who walked into a store would see the same choices and prices that everyone else saw. Of course, the retailing world wasn't nearly as democratic as it claimed to be. In ethnically and racially divided cities, different neighborhoods often saw very different products and prices. And certainly, small merchants with personalized approaches to clients continued to exist. But mass equality was the ideal, and personalization was pushed to the side.[21]

Reaching huge populations was also the goal of the twentieth century's rising advertising industry. Advertising itself goes back millennia; some call it the second oldest profession. For most of that time, it was a handicraft activity practiced by the creator of the product to be advertised, or by someone that person had hired to write handbills or announcements placed in narrowly circulated newspapers, books, and magazines. This changed during the nineteenth-century Industrial Revolution, when what we know as the advertising industry was born. During this time, new, routinized approaches to production led factories to turn out all kinds of products previously made by hand in huge numbers and at much lower prices than before—as well as novel products like toothpaste, corn flakes, safety razors, and cameras. The large number of items encouraged competition among manufacturers of similar goods, who turned to advertisers to increase sales. These advertisers, who often looked to the immigrant-swollen cities with their many stores for new customers for their clients, placed ads in magazines like the *Saturday Evening Post* and newspapers like *New York World,* which reached tens of thousands, or even hundreds of thousands, of middle- and upper-middle-class people. The periodicals, in turn, charged readers little (typically the price of delivery) in order to gather the huge audiences that advertisers were paying to reach. Commercial radio, beginning in the 1920s, and commercial television, beginning in the late 1940s, initially followed the same model. In their cases, no-fee access by the public—buy the radio, plug it in, turn the dial—was supported financially by advertisers seeking "mass" audiences.[22]

Despite mainstream sponsors' focus on the mass market, with audiences as large as possible, some businesses thrived by pursuing much smaller slivers of the sales arena. A small but important area involved personalization, in the peddlers' sense of learning as much as possible about an individual, though the firms did it through investigations rather than personal interactions. Prominent among these were credit bureaus, which arose in the late nineteenth century to help merchants figure out how risky it would be to deal with people who wanted to take home a product and pay for it over time. Although some of these firms "produced little more than blacklists of debtors and delinquents," others developed ratings systems and reference books that contained estimates of thousands of individuals' ability to pay their bills.[23] These bureaus proliferated from the first half of the twentieth century until the 1980s, when a few firms, having leveraged the efficiency of computers to collect and store credit information, bought out all their competitors. The three national bureaus that eventually emerged from that consolidation owned databases with personalized files for more than a hundred million individuals.[24] Unlike medieval peddlers, these giant credit bureaus had no relationships with the people they were scrutinizing. Still, people who received low credit scores often reacted by finding out what they could do to raise their monetary reputations.

Other marketing and media businesses also focused on particular segments of the mass market, but they weren't nearly as individualized as credit bureaus when it came to describing or interacting with their target audiences. The technologies of printing and broadcasting didn't allow them to personalize the creation or distribution of their articles or programs. Instead, they created specialized publications for demographic groups they had identified as large enough to be monetarily useful—such as farmers, college students, immigrants speaking different languages, African Americans, religious groups, and the wealthy—and use them to advertise products that were popular with those demographic groups, or that those groups in particular needed. In a mutually reinforcing process, magazines, newspapers, local radio stations, and local television stations began to survey their audiences to learn their buying habits; they then tailored their content to attract the most advertisers. In the decades after World War II, marketers and media firms began to divide groups of people not just according to demographic categories like gender, income, race, or age, but also by geographic location; lifestyle preferences such as hunting, skiing, owning a pet, or being a food connoisseur; and political and cultural values. "Segmentation" became a buzzword among marketers and media firms.

The proliferation of cable television channels in the 1980s pushed segmentation into high gear. Media firms sold their audiences to advertisers in chunks. Success depended on the number of people in the targeted segments who bought an issue, viewed a program, or passed a billboard. The interaction between the advertisers and audiences went almost entirely in one direction. Measures of actual audience interactions with the ads were rare and very delayed. Instead, media and advertising firms typically relied on various forms of survey research (such as the Nielsen company's familiar ratings of TV viewership) to measure their success. Sometimes the demographic, even psychographic, segments that marketers identified were large enough and attractive enough that firms didn't just advertise differently to them but also created variations of popular products—brand extensions—specifically for them. "In the old days," a Procter & Gamble executive noted in 1994, "Tide was one big brand. It stood for clean, white clothes, and all women 18 to 49, whether they had kids or didn't have kids, washed their clothes [with it]. But now, you have Tide with Bleach, Tide Ultra, Tide Unscented. And each of these brands are still targeted at women 18 to 49, but they are targeted at differences between segments of women 18 to 49."[25]

Little did this Procter & Gamble executive know that the same year he spoke so confidently about segmentation, 1994, the commercial internet would emerge with the development of the popular Mosaic web browser. Following a 1993 siren call in *AdWeek* that "the internet is opening up to tens of millions of personal computers," companies invested in the web began to move advertisers away from segmentation toward the first widespread push toward marketing personalization since the age of itinerant peddlers.[26] Using computers to tailor many different messages to different individuals became efficient in ways that traditional media could never achieve. Many in the industry thought of the shift from segmentation to personalization as an idealized, electronic vision of the traditional salesperson's ability to learn just what a customer wanted through one-to-one interactions.

At first, personalization used many of the same social categories as segmentation, but it increasingly allowed marketers to link those aspects of individuals' backgrounds and lifestyles to specific behaviors both on and off the internet. The capability started with the internet industry's creation of cookies and other tags, which allowed publishers (websites and then apps) to follow, record, and store individuals' movements on their sites and elsewhere in digital space. By getting users to log in with email addresses and passwords, publishers could learn their visitors' names and addresses and then purchase more information about their gender, age, income, and purchasing habits from data brokers, who themselves used the internet as a feeding ground for data. The publishers offered marketers the ability to reach the specific people they cared about or who had the alignment of characteristics that the marketers wanted. Moreover, media firms auctioned off to advertisers, often in real time, those individuals using the web, apps, and increasingly digital television and radio who had particular interests and buying power, and who were logging on from specific locations—for example, working mothers who can afford a luxury SUV, seem to be in the market for a new vehicle, and are at an airport. Immediate measures of the individuals' interactions with the ads and with the content around them became common. Moreover, the commercial messages—and sometimes the actual items—that marketers showed individual users grew increasingly personalized. Publishers themselves followed this approach. Two people visiting the same site at the same time could see different goods, stories, prices, and discount coupons, depending on the publishers' profiles of them. Even though research showed that Americans don't like the idea, and even though the algorithms that drove the conclusions could be based on prejudicial samples or assumptions, these practices were and are completely legal.[27] Beyond certain areas of health and financial information, few nationwide regulations govern the gathering, exchange, and use of data about people in the digital realm. The Federal Trade Commission's response so far has been to encourage companies that use people's data to self-regulate around the principles of giving people notice and some measure of choice, even if the choice is only to leave the site or app.

In recent decades, a major industry has grown up around personalization. The momentum was such that a 2019 report from the Gartner consulting firm called personalization "a top priority for application leaders working on digital commerce and customer experience." Gartner's all-encompassing conception of the process goes beyond linking individuals to the traditional segmentation categories to include such things as where the individuals are when the marketer is communicating with them, what products they like to buy with other products, favorite brands, frequented events, most-liked sales associates, and most-used products. Gartner suggests that knowing individuals' interests and hobbies, family relationships, opinions and attitudes, their friends on social media such as Facebook or Instagram, and even which of their devices at home are connected to the internet—smart speakers, thermostats, lamps—can help generate a comprehensive profile of a customer's life. "All this information," the report notes, "can be captured in a CRM [customer relationship management] system, where a detailed customer profile can be maintained to enable a 360-degree view." Of particular importance, according to the report, is an ability to reach the same individuals across all the media outlets they visit. Sometimes the data can be used for advertising to particular groups of people— for example, women with specific cooking-related hobbies. But increasingly the information is used to help the company advertise to the profile created around a particular person, or what Gartner calls that individual's "personas and motivators."[28]

Marketing experts often use the term *big data* to describe the high volume, velocity, and variety of information they collect to make personalized inferences about individuals and predict their actions. This new level of specificity means the strategy of dividing the audience into chunks and then marketing to the different chunks has become increasingly obsolete. "Data science is upending the idea of segmentation," said the director of an analytics lab in 2017.[29] His goal is instead to crunch many bits of information regarding individual shoppers, forecast their behavior, and set up structures to sell to them efficiently. This method typically requires complex mathematical tools called predictive analytics. Specialists in this area use data mining, logistic regression, and cloud computing to track people's everyday activities via "the little data breadcrumbs that you leave behind you as you move around the world," in the words of MIT computer science professor Alex Pentland. Catalina Marketing, a company that follows people's purchases at store checkouts, used a different metaphor: "Like fingerprints," the firm wrote in an online 2015 brochure, "every shopper's profile is unique in the assortment of products they buy each year." Using information about these trails of purchases, and increasingly complex models, marketers and publishers created profiles of individuals that scored their value to a merchant, calculated the chances they would click on particular ads, and predicted whether it would be worth showing them a specific product.[30] This puts to commercial use Eric Schmidt's comment, when he was Google's CEO, that the information Google collects on individuals enables it to know "roughly who you are, roughly what you care about, roughly who your friends are"—to the extent that it knows its customers better than they know themselves.[31]

The glitch in this one-to-one nirvana is revealed by Schmidt's use of the word "roughly." The aim of the personalization drive is to surpass any customer smarts that an old-fashioned salesperson could develop, but the collection of data points about people and their behavior that marketers bring together is sometimes wildly inaccurate. The secret worry of internet marketing—a view that burst into the open during the 2010s—is that tracking, profiling, and targeting are fraught with challenges for the entire digital ecosystem. Data may not be up to date, profiles may be created based on multiple users of a computer or phone, names may be confused, and people may lie about their age, income, or even gender in order to confuse digital marketers.

Advertisers are also uneasy with the well-known problems of click fraud and ad blocking. "Click fraud" takes place in pay-per-click advertising when the owners of digital locales that post ads on a domain are paid according to the number of visitors to those domains who click on the ads. The fraud happens when a site or app owner pays people, or creates an automated script, to click on ads to accumulate money deceptively.[32] "Ad blocking" is using software (less commonly, computer hardware) to remove advertising content from a webpage or app.[33] While industry players debate the specific reasons for these activities and have tried a variety of solutions, they agree that they cause substantial economic losses.[34] A 2018 study by Adobe found that as much as 28 percent of all website ad traffic was click fraud.[35] A 2017 report by the Juniper consultancy projected just 9 percent for 2018, but even that lower proportion would cost the industry about $19 billion despite major attempts to combat it.[36] Juniper estimated that by 2022, ad fraud would cost the industry $44 billion annually. As for ad blocking, eMarketer found in 2018 that one-quarter of all U.S. internet users—71 million people—block ads on at least one device on a regular basis. The firm added that this "share will continue to rise as consumers express frustration with their digital advertising experiences."[37]

Confronting all these problems in a November 2019 report, Gartner noted that marketers' efforts at personalization were also being frustrated by a "continuing decline in consumer trust" and "increased scrutiny by regulators." Gartner's head marketing analyst explained that "consumers have developed an increasingly jaundiced eye toward marketers' efforts to embrace them" and that "their increasingly cluttered email inboxes and mobile phone notification centers may lead them to ignore even the most carefully personalized and contextualized message." Perhaps more important, Gartner's survey of marketers showed that roughly a quarter believed that weaknesses in data collection, integration, and protection are obstacles to personalization. The company predicted that by 2025, "80% of marketers who have invested in personalization will abandon their efforts due to lack of ROI [return on investment], the perils of data management, or both."[38]

This odd conclusion contradicts the consultancy's own writings. Just months earlier, Gartner had heralded the importance of tailored communication. "Customers expect to be recognized and want their experiences personalized," stated a "Smarter with Gartner" online essay in April 2019. It suggested that tensions around data use could be eased by being upfront with customers about what data would be used, where, and in what context. (That suggestion would also fit with increasing government demands for data-use transparency.) But its overriding point was "organizations that combine identity data with behavioral data will outpace those that don't."[39] Even the November 2019 report that predicted personalization's demise exhorted marketers to move forward with it, only with more care than previously and with greater input from various groups within their firms. "Collaborate with cross-functional teams to align personalization efforts and increase momentum," the firm advised. "Sharing control of personalization efforts can lead to shared insight and expand collective impact and ROI." The same report forecast that "by 2024, artificial intelligence identification of emotions will influence more than half of the online advertisements you see. . . . Companies like Amazon, IBM and Walmart envision a future where they combine biosensors and AEI [artificial emotional intelligence] to detect emotions to influence buying decisions."[40] By definition, using AI to identify individuals' emotions and so decide which online ads to show them is a form of personalization.

Rather than dismissing personalization, then, Gartner's list of concerns about it is connected to and explains the never-ending personalization cycle. The marketers I interviewed for this book certainly didn't discuss voice intelligence as a substitute for tracking known people on the web, on apps, in stores, and everywhere in between. In the words of Pete Erickson, whose company creates conferences for voice-tech developers, voice is a "value added" to the current personalization regime, not a replacement.[41] But marketers do see the rise of voice as a new beginning, an opportunity for their data scientists to use artificial intelligence to achieve unprecedented and near-immediate insights into shoppers' identities and inclinations without the fraud and other problems with "traditional" personalization. To people at the leading edge of voice developments, the public's embrace of smart speakers, intelligent car displays, and voice-friendly phones, along with the rise of voice intelligence in call centers, presents an intriguing way to get around the current difficulties of knowing an audience on the internet. Not only can individuals be profiled by the words they choose and by the systematic relationships among them (their *speech patterns*), they can also be assessed by the physiology of their sounds (their *voice*), which according to some researchers is unique and cannot lie.

The hope for a new chance at audience knowledge is where the furious race for the next cycle of personalization begins. Currently, using voice to infer emotions is the centerpiece of this emerging phase. "Reading human emotions and then adapting consumer experiences to these emotions in real time" will "help to transform the face of marketing," says an executive from Affectiva, a consultancy spun off from MIT.[42] Executives in the call services business also see voice-driven personalization as marking a new era, with emotions analytics at its heart. The CEO of Clarabridge, a provider of phone response technology, foresees that new voice technologies will help identify "key indicators" of a particular caller's loyalty "such as effort, emotion, sentiment and intent." The head of another firm asserts that "this drive to personalization is already benefitting from using artificial intelligence to learn deeper meanings in what people say."[43]

Companies large and small are rushing to shape this next frontier of personalization. To do it they are creating a new agenda for mining customers' voices using computer-driven artificial intelligence processes, one that harks back to the work of one-to-one salespeople but without the give and take, and with the customer having only the dimmest understanding of what is taking place. The hallmark of this new world is what we might call the *industrial construction of audiences*. Our ideas of who we are will come from the efforts of many organizations working together. Some will provide software, others hardware, others data about individuals—all to create personalized profiling that will predict our commercial value in any given situation. The guiding principle is that the new gold in people's speech is not in what they say but how they say it, and how they sound when they say it. A common way of getting to the nuggets involves machine learning based on samples called training sets. Take two labeled training sets—say, one of people who are nervous and one of people who aren't—and feed the data to an algorithm, a dynamic formula that is designed to detect patterns among a large number of data points about speech: tones, speed, emphases, pauses, and much more. Some of these speech characteristics are so complex that only a computer with particular instructions could detect them. Say the algorithm learns to pick up the subtle signs that indicate (at a certain level of statistical confidence) whether a speaker is part of the nervous or non-nervous group, and that it can do the same with new samples in the future. A next step might be to determine whether the nervous-sounding group is more likely to buy certain products, default on loans, or do other things that marketers would find important. Still further steps might involve linking those findings to other knowledge about the individuals. It's possible that adding information about gender, geography, and jobs, for example, could add to the ability to predict whether a nervous person is more likely to buy certain foods, take certain vacations, or return products often.

There are other, more complex versions of machine learning. One family of methods, deep learning, uses algorithms in multiple layers that try to find patterns in different aspects of the phenomenon (say one layer concentrates on the waveband frequencies of the voice and another on the ways syllables are used); these patterns are then analyzed together to draw conclusions about the whole. Optimists about voice profiling point to attempts by healthcare providers to detect diseases such as Parkinson's or post-traumatic stress disorder. Characteristics of a person's voice may even indicate suicide risk, according to a team led by Louis-Philippe Morency, a computer scientist at Carnegie Mellon University.[44] Morency's group tentatively concluded that if people who have attempted suicide have "a soft, breathy voice," they are more likely to reattempt than those "with tense or angry voices."[45] Such efforts are intriguing, but scientists are at the beginning of a long road, with many questions ahead about the accuracy of their results and whether factors like the instruments they are using or other circumstances of testing, rather than the voice itself, could be influencing their findings. Another concern is the confidence level: how statistically significant should the findings be before they activate social programs—for example, to check the voices of people who have attempted suicide? Should 90 percent confidence be enough for getting a voiceprint and potentially disrupting someone's life, or should 99 percent be required? And lurking behind all the results is a worry about algorithmic bias: that if the training sets are not large or diverse enough, the conclusions reached might not reflect the entire population and may even lead to discrimination. Consider, for example, if the training set of suicide attempters didn't include people whose voices have always been breathy, regardless of their mental state. If a social policy based on voice tagged them as suicide attempt repeaters, it might cause unnecessary trauma for them and their loved ones. These sorts of discrimination often relate indirectly to race and gender. The University of Washington's Ryan Calo is one of many observers who argue that machine learning inevitably introduces race and gender biases.[46]

Despite these cautions, businesses—starting with customer contact centers—have been moving confidently into voice intelligence. "Today we're able to generate a complete personality profile," said the CEO of Voicesense, which claims to be able to use people's voiceprints to accurately predict loan defaults, people's likelihood of filing insurance claims, and customers' investment styles, among other key indicators. The head of the CallMiner call analytics firm was just as bold. "Our vision is to empower organizations to extract meaningful and actionable intelligence from their customer conversations," he stated. "AI has become a cornerstone of providing those capabilities with efficiency and scale." The CEO of voice-emotion detection startup Beyond Verbal summed up the claim and implied the ambition. "Nearly two decades of research," he asserted, make it clear that "it's not what someone says, but how they say it, that tells the full story."[47]

You may be thinking that personalization by marketers makes many people nervous. Heightened concerns about privacy and data losses speckle the news and pepper the talk of lawmakers. Wouldn't an effort to push voice intelligence on American society cause a furor, even if the industry doesn't show all its cards? At this point few people know that when they call 800 numbers for service there is a decent chance that how they are treated will be partly based on a computer analysis of their voice. But eventually the industry's cover will be blown. So it makes sense that the voice intelligence executives at Amazon, Google, Apple, Samsung, Bank of America, and other companies should already be thinking about this nervous environment. They surely know, for example, about Gartner's 2018 finding, reported in the *Wall Street Journal*, that 63 percent of four thousand people surveyed in the United Kingdom and United States "didn't want AI to be constantly listening to get to know them."[48]

That's where *seductive surveillance* enters the picture. Pinelopi Troullinou coined the term in her 2017 doctoral dissertation to explain why people "willingly" participate in activities that allow organizations such as cell phone companies to keep tabs on them.[49] The phrase describes an effort to present target audiences with packages of devices, prices, and possibilities that are attractive enough to overcome any concern they might feel about the marketing surveillance carried out through those devices. And since surveillance drives personalization, companies with an interest in personalized inferences based on voice want to create an environment that keeps people buying and operating the devices despite any gnawing surveillance concerns.

It appears the voice executives understand that the best way to do this is to turn people's use of voice assistant devices into a widely accepted habit. To academics this is a familiar approach. The twentieth-century French sociologist Pierre Bourdieu famously used the term *habitus* to describe a person's mindset regarding the carrying out of routine activities—that is, habits. He asked, "How can behavior be regulated without being the product of obedience to rules?" His answer was that regularized behavior results from a person's interactions with social position ("capital") and the goings on in society ("the field"). These forces are not static; changes in a person's social position and in what governments or companies do can affect the person's mindset in particular ways. Bourdieu's key point is that a person's mindset is both shaped by outside forces ("the field structures the habitus"), and once internalized, helps a person understand and accept the logic of the institutional forces that created it ("habitus contributes to constituting the field as a meaningful world").[50]

The British sociologists Tony Bennett and Francis Dodsworth believe that Bourdieu's concept of habitus is useful in understanding how people become socialized into acting predictably in the world, but that a major limitation in his writings is that he focuses on the individual's habitus—what and how people think systematically about the world—while paying little attention to the ways companies, governments, and other "material agencies" shape the mindset and the habits that flow from it. Bennett and Dodsworth argue that an understanding of "the processes through which habits are formed and reformed" must take into account their relation to a wide range of material things in society, from the effect of schoolroom desks on young students' routines to the ways that specific logistical models in a company influence how employees handle shipping. They use the term *habituation* to refer to the process by which forces in society cultivate the creation of habits.[51]

It's not hard to see how habituation and seductive surveillance are linked. Seductive surveillance is a dual strategy—a balancing act—by which companies get people in the habit of using voice across a range of devices. Troullinou suggests how the seduction works in ways that allow organizations to keep tabs on individuals. With phones, she writes, "the user is seduced by discourses of convenience, efficiency and entertainment into handing over personal data, and thus being transformed into a subject of surveillance."[52] She quotes two researchers who say that designing a seductive product "involves a promise and a connection with the audience or users' goals and emotions" and suggests that "seduction operates at multiple levels, from technology to marketing discourses and governance."[53] Though she doesn't specifically mention the phone's voice assistants, the generalization would presumably apply to them, too.

In his book on "emotional AI," Andrew McStay addresses the practical reasons that voice intelligence devices like the Amazon Echo attract users. For one thing, they are "an increasingly popular way of using the internet and engaging with on-demand assistants." He also notes that "branded voice assistants facilitate *intimacy at scale*, which means personalized conversations whenever and wherever required, no wait-times for customers."[54] Bret Kinsella, a voice industry consultant, observes that the devices "are designed to help simplify users' lives. Over time, more and more agency will be granted to voice assistants to simply execute tasks on behalf of the user." Consumers "will not necessarily care how the task is fulfilled, just that it gets done." Some of this will be directed by the users, such as when Google Assistant makes a restaurant reservation, but some will be proactive; for example, "a voice assistant will notice a favorite item that the user regularly purchases is available on discount and it will simply be ordered and shipped without any explicit instruction."[55]

As we will see, Amazon, Google, and other firms don't rely on the hope that people will immediately see the value of their devices and rush out to buy them. Nevertheless, this focus on seduction provides a useful beginning for exploring how companies persuade people to regularly speak to devices even as they

are uneasy about the implications. The academic and trade literature says much less about how firms play down surveillance in
order to reduce users' concerns. Communication professor James
Katz suggests that "over time people get inured" to being tracked
and profiled.[56] It's an important insight, but it doesn't explain
why people didn't resist actively before they became inured. One
answer is that from the start, companies try to stop people from
learning what's taking place. This is a refrain that shows up often
when researchers explore how companies with other types of
technology relate to the public. For example, Sarah Roberts,
writing about social media firms, suggests that the platforms
cultivate an "operating logic of opacity" that discourages users
from trying to engage with these systems.[57] Nora Draper and I
have written about the features of language that internet firms
use to cover up the surveillance they are enabling.[58] One common
approach is placation, where a firm appeases customers by
assuring them it cares about their privacy—often in ways that
bear no relation to what is described in the privacy policy. Another
approach is diversion, or trying to get customers not to pay attention to disclosures about information use. You know about this
approach if you've ever tried to find a website or app's privacy
policy; it's in tiny letters at the bottom of the page or way down
in choices on the app. If you actually attempt to read the policy,
you'll find that it's often full of jargon, as if it were deliberately
written to confuse and discourage the reader.

Draper and I argue that these routine corporate practices not
only obscure what the companies are doing; over time, and as
habituation develops, they encourage a sense of *resignation*—a
feeling among users that even though they would like profiling

not to happen, they can't do anything about it. Such feelings of resignation showed up in research by a team from the Annenberg School for Communication in 2017 and 2018. From a representative telephone (cell and wireline) survey of the U.S. population, we characterized 58 percent of respondents in 2017 and 63 percent in 2018 as resigned based on their agreement with two statements: "I want control over what marketers can learn about me online" and "I've come to accept that I have little control over what marketers can learn about me online." We also explored Americans' sense of resignation—as opposed to whether they believe in trading data for benefits in a calculated way—through questions about supermarket shopping and loyalty programs. We found, for example, that a large proportion of Americans—43 percent—say they would let supermarkets collect data about them despite indications elsewhere in the survey that they disagree with consumer surveillance. We also found that the more they know about the laws and practices of digital marketing, the more likely Americans are to be resigned.[59]

Following our report, a handful of other studies identified related sentiments, reinforcing our finding that many people are resigned to companies' use of individual data.[60] As we will see, the voice intelligence industry appears to cultivate resignation as a way to play down press concerns about surveillance. The industry's recurring theme is that giving up your voice for activities in and out of the home is inexpensive, convenient, fun, emotionally satisfying, and natural—even if it can make you feel nervous and unable to control information about yourself. Voice technologies are also touted as offering increased choice and personal efficiency—the very definition of individual sovereignty in modern times.

The voice assistant's central purpose as a marketing tool, however, complicates this idea of personal freedom in alarming ways. By design, buying into voice-intelligence technologies seduces you to relinquish the sovereignty of this part of your body so that companies can use it to audit you, assess your value, and perhaps discriminate against you in ways you may never learn of or understand. As more and more of us rely on voice assistants and other voice-activated technologies, the data that we provide, and that companies mine, mean that "freedom" may well become the corporate presentation of choices personalized for you based on the proposition that this is what you want because your voice doesn't lie. When this process is multiplied by hundreds of millions of people encountering billions of such personalized choices, we risk ending up with a society that is habituated, or resigned, to equating freedom with its opposite—biometrically driven predestination—in all areas of life.

1 RISE OF THE SEDUCTIVE ASSISTANTS

Amazon today quietly unveiled a new product dubbed Amazon Echo. The $200 device appears to be a voice-activated wireless speaker that can answer your questions, offer updates on what's going on in the world, and of course play music. Echo is currently available for purchase via an invite-only system. If you have Amazon Prime, however, you can get it for $100. . . . Amazon wants to bring the digital assistant to the living room. The idea is a very interesting one, but it's difficult to imagine there being a lot of demand. Given that many of these features are already offered in mobile devices, most users will be happy to continue getting updates to their assistants there. Then again, Google Now, Siri, and Cortana are far from perfect, so Amazon does have some wiggle room. We'll have to reserve further judgment until we can get our hands on one.[1]

These comments began and ended a rather short VentureBeat article published on Alexa's launch day, November 6, 2014. The piece was obviously correct that Amazon wanted to copy the success of the digital assistant, the voice-enabled phone helper championed in the United States by Apple, Google, and Microsoft Cortana. In a video introducing the Echo, Amazon portrayed the living room and kitchen as landing spots for its hands-free assistant, Alexa. In subsequent years Amazon would try to colonize the entire home with Echo devices. That goal was made very clear in the launch-day video, which featured testimonials by customers who had tested the product before its release. Among the uses they excitedly mentioned were finding out the weather, helping with recipe measurements, learning new jokes, reading books, helping a blind couple to set timers, playing the news, and compiling shopping lists. "The Echo," one customer said, "is a tool we use to keep our household functioning."[2]

The video also shows that right from the start, Amazon used a strategy of seductive surveillance. It presented biometric identification and profiling as part of the device's features. The stories in the video demonstrate how the Echo recognizes individuals by their voices; in one home, Alexa learns to understand a man's English despite his German accent. Any concerns about this level of knowledge are quietly swept aside by users' enthusiasm for the device; we're encouraged to see only the benefits of talking to our own Echo, which in turn can hear and remember each of us. To further set the hook, Amazon offered an early-purchase discount to its most trusting customers, the Prime members whose loyalty to the company had earned them free shipping and other benefits. Such discounts would become key parts of Amazon's long-term seductive surveillance strategy.

Despite the company's exuberance, some articles belittled the new stationary assistant. James O'Toole with CNN Business, for example, commented that "Amazon's quirky Echo is Siri in a speaker" and that "this may be another case of a product that you can render superfluous by simply taking your phone out of your pocket."[3] Others, though, marveled at Amazon's boldness in entering a technology realm that was both mind-bogglingly complex and already filled with competition. Still others didn't seem to get it. CNET dutifully ran a story, headlined "Amazon Debuts Siri-like Digital Assistant Echo for Your Home," but—perhaps indicating the writers' low engagement with the product—neglected to mention that Alexa would use artificial intelligence to interact with family members and would assess and retain what they asked about a wide range of topics. Also missing from the announcement: that Alexa's setup app would ask where in the home the Echo was placed and would request that each user create an identifying voiceprint. What these individuals asked the intelligent speaker, as well as how they asked it and in what room, gave Amazon information with which to create profiles of the family's needs and concerns in the highly personal environment of their own home. By using artificial intelligence more intensively than previous assistants, Echo could give marketers access to an environment they had never been able to penetrate directly.

When Alexa was launched, Amazon was already applying the latest computer analyses to profile people on its website, on its advertising network of other sites, and on apps. It knew who its shoppers were, what they were like, and often what they were doing on the internet. The company acknowledged using profiles to understand the buying patterns of various population groups and to tailor the product choices and ads on its site, apps, and ad network to what it had learned about individual users. Yet I have never been able to find a public statement from Amazon about how it intended to use the new storehouse of information the Echo would provide: what individuals said to Alexa, and how and where they said it. Nor did the firm disclose how it would tie the knowledge it gained about individuals from the intelligent agent to the profiles it continued to assemble by other means. What does seem clear is that the company didn't want people worrying about the new flood of data that Alexa would send its way. It was no accident that the company worked to create strong personal bonds between humans and its humanoid, so that customers would happily allow Alexa onto devices not only around the home, but also in the car, in hotels, in stores—everywhere. Part of the seductive surveillance strategy was to position Alexa, with its soft female voice, as a helpmate rather than as an inquisitive salesperson.

In selling the friendly comfort of a female virtual voice assistant, Amazon was following the paths charted not only by Apple, Microsoft, and Google, but also by the contact center business, which handles customer-service inquiries for a wide range of companies, and for a range of purposes. At this point, contact centers were leading the way in using voice data acquired during customer-service calls to categorize and persuade callers with as little human labor as possible. By the late 2010s, the aims of the contact-center business and those of the intelligent-assistant business had begun to merge. Both were convinced that the sound of a person's voice had value in the marketplace. Both privileged computers over humans in drawing inferences about people's speech and voice patterns and in building "satisfying" relationships with customers. And both pushed technologies that could dig deeply into customers' private interests by combining more traditional marketing-related information like age, gender, income, race, lifestyle, and online behavior with data about what they were saying, analyzed in ways the customers would hardly notice or understand.

The technological breakthroughs that led to Alexa were a long time coming. The earliest step was the basic effort to replicate the human voice—which as it turns out is no easy feat. As early as 1773 a German-Danish scientist named Christian Kratzenstein created models of the human vocal tract that could produce vowel sounds.[4] But it took more than a century of additional attempts before Thomas Edison invented in 1877 what was to become the first marketable device to record and play back voices and other sounds.[5] The next ninety years involved a slow process of creating machines that could either synthesize or recognize spoken words, but not both. Only toward the end of the twentieth century did engineers begin to develop speech-synthesis systems that could interact flexibly with humans.

In the United States, the business of voice intelligence pushed forward with the support of both taxpayer and private money. The taxpayer funds came from the U.S. Defense Department's futuristic investment arm, the Defense Advanced Research Projects Agency (DARPA), apparently with the goal of developing the role of voice on the battlefield. In 1971 DARPA's Speech

Understanding Research program funded five years of university research toward creating a machine that could understand at least a thousand words. The greatest success was Harpy, from Carnegie Mellon University, a computer capable of understanding 1,101 words. While linguistic and engineering knowledge played a role, much of the increase in the number of words understood had to do with a growth in computer capabilities that would take off over the next decades. In 1976, as DARPA's first speech funding program ended, the best computer available to researchers might need a hundred minutes to decode just thirty seconds of speech. As computer processing speeds and memory grew, so did word understanding. By 1990, a typical commercial speech recognition system could handle more words than are in the average human vocabulary.[6]

Even more consequential during this period were theories developed by scholars at universities and in private firms about what it means to recognize speech and understand it, sometimes to the point of being able to pick out one person's particular vocabulary. Success came in halting steps. A scientist on IBM's speech recognition team during the 1980s recalled that their system, which required a roomful of computers, was "trained" to understand only what a particular individual said. If the computer made only one error for every ten words, that was a terrific result.[7] Over the following decades, investigators around the world, especially at IBM and Bell Laboratories, created artificial intelligence algorithms that improved computers' abilities to understand human speech. As the market research company Forrester notes, "the killer feature of AI algorithms is their ability to learn the underlying patterns in any phenomenon, regardless of complexity, given enough relevant data and computing power."[8]

The key processes involved in Alexa are speech recognition, speech processing, and speech creation or synthesis. Each step requires large and varied datasets of recorded and transcribed speech to train the system. In the training related to speech recognition and processing, engineers use complex statistical models under the rubric of machine learning (which nowadays involves powerful tools called deep learning and deep neural networks) to teach the computer how to link sounds to words and sentences so that it will transcribe them correctly, irrespective of accent. Once the words are transcribed properly, the goal is to use a set of statistical procedures called natural language processing to understand the meaning of the speech—what the person is trying to say. To do that, engineers again use large and varied training sets. The goal is for the computer to interpret the statement correctly and take the correct action. Although the sentence "Wake me up at 7 a.m. tomorrow" seems simple, an assistant would have to know that several variations on this request—for example, "Set an alarm for 7 a.m.," or "Please wake me at 7 a.m."—should yield the same result. A good training set allows the deep-learning algorithms to see that a large variety of such statements should yield the same output.

Then there is the matter of training the computer to respond, also by voice, which involves a series of extraordinarily complex steps. The process often works this way: first, the engineers find a professional voice talent whose sound meshes with the creator's aim, including the personality the creator wants to give the assistant in the language being used. That leads to recording sessions: ten to twenty hours of speech in a professional studio. The actor's scripts, according to Apple's Siri team, "vary from audio books to navigation instructions, and from prompted answers to witty jokes."[9] The engineers then run the words spoken during those sessions through a computer that slices them into their elementary components, their snippets of sound. The computer will use a database of these speech snippets when it needs words that sound certain ways. The final step, recombining the snippets into sounds to match the sentences in a text, is the hardest part. The current approach is to use deep-learning methods on the training set (which links audio and transcribed sentences) so that the training set will teach the text-to-speech computer how the words in a text ought to sound. This means using acoustic models to give the computer the probabilities and other data resources it needs to decide how to choose and link snippets to convey not just the words, but also a wide range of emotions through voice tonalities, rhythms, and cadences.

This is only a very basic sketch of the astonishingly complex set of decisions that a computer assistant makes in responding to an apparently simple command or question. Sometimes the assistant may take a shortcut by focusing on specific keywords in a sentence—for example, *what* and *time* in "What time is it?" At the same time, engineers are working hard, and with increased success, to understand the multiple kinds of context surrounding what people say. The most basic is contextual understanding: when a person says "set an alarm," the assistant responds, "what time do you want?" Other contexts might involve understanding the person's remarks differently depending on geographical location, room in the home, time of day, or even, for a smart watch, the person's pulse rate.

Although it was clear early on that personalization using voice intelligence could be enormously valuable, the progress of artificial intelligence in this area was by no means smooth. Vlad Sejnoha, a computer scientist who worked for Nuance, told me the company grew by making strategic investments during what people in the voice analytics business call the "speech technology winter" of the 1990s and early 2000s. "There had been a number of notable failures in the late 90s," he recalled.

> Companies overreached; the technology was really not up to what they were trying to accomplish. . . . The computation wasn't right there, the connectivities weren't quite there in the 90s. PCs weren't really all that powerful. And so a lot of the applications that were available were clunky and certainly underperforming especially compared to today's standards, where in many cases cloud-based speech recognition just works. It's reliable and accurate for the great majority of the population. That was not the case [back then]. You had to laboriously train several recognition systems. For example, if [a customer] bought Dragon Dictate in the 90s, you [had] to spend a couple of hours training it, and it was an expensive product. So these companies ran into trouble. . . . And a lot of large companies, including Google, Amazon, and Microsoft and Apple minimized their investments, if they had any.[10]

Sejnoha recalled that Nuance's CEO at the time, Paul Ritchie, "had a lot of foresight and used that time to accumulate a lot of speech technology assets, and made a lot of acquisitions early on. He was investing for the time that he and the rest of us believed would come again, and indeed it did. And I think there was a time Nuance stole a march on a lot of these giants, and some of the early products I think caught Google and Microsoft and others by surprise." In the mid-2000s, "they quickly started investing again, and it's well known—it's a matter of public record—that in many cases they [did] that using licenses from Nuance." By the late 2000s, computer speech recognition and appropriate responses had advanced enough that a Microsoft executive used it to schedule his appointments, and a Microsoft lab was trying out a "medical avatar" that could ask children questions about their symptoms and make diagnoses based on their answers.[11]

While these trials were taking place, marketers were beginning to apply this growing area of artificial intelligence to a crucial but controversial part of their business, the customer contact center. At the start of the twenty-first century, contact centers arguably had access to more information on Americans than any other marketing endeavor, but they struggled to use the data efficiently to personalize interactions. The basic problem was an old one: the call center, as it was originally called, was about a hundred years old. Big department stores had created the first ones, which were simply large switchboards. Wanamaker's department store in Philadelphia established the first store telephone system around 1900, twenty-four years after Alexander Graham Bell first exhibited his invention. By 1915 the store had the largest private branch telephone exchange in the world, with more than two thousand operators who handled over 1.8 million messages.[12] American Telephone and Telegraph (AT&T), the phone company, had an operator pool that made millions of verbal contacts with customers each day. But those two firms were giants of their day; many marketers were unwilling to invest in the kind of response infrastructure that Wanamaker and AT&T created. Gradually, a call center industry evolved, consisting of companies that handled phone calls for multiple clients. People in the industry remember its early years as filled with human error, unreliable technology, and slow service. As one history of the business notes, "back in the day, holding for 10, 15, or even 30 minutes wasn't unheard of."[13] The goal was just to keep up with the flow. Harried phone agents inevitably made judgments about callers based on how they spoke, but their conclusions weren't recorded; they just wanted to complete the call.

Little changed in how the centers dealt with customers until the 1960s. That was the threshold for several decades of developments that both sped up call handling and gave the call industry far more information about customers than any other media business could obtain. Ironically, all the developments started a long-term movement by call centers away from the human salesperson's intuition about voice, toward judgments based on hardware and software. In the 1960s, AT&T introduced toll-free 800 numbers and began to replace rotary dials with touchtone calling. The 1970s brought automatic call distribution (ACD) systems and interactive voice response (IVR). Toll-free numbers were a revolutionary marketing innovation at a time when long distance phone calls could be expensive. Accompanying advertised products in print media and on television, the numbers allowed people to buy things over the phone with their credit cards (or cash on delivery) and have them mailed to their homes. Automatic call distribution replaced manual switchboards with computer-guided ones that could allocate the new torrent of toll-free calls to operators far more efficiently than would have been possible in previous decades. Further increasing routing efficiency, the interactive voice-response systems played digitized speech messages to callers before they reached a live person and instructed them to push one or another touch-tone button to indicate the purpose of their call. That way the automatic call distribution would not only route the calls to a waiting representative; it would also put callers in touch with a representative who had the skills the caller wanted and who knew the basic reason for the call.[14] It was the start of automated personalization.

During the 1980s and 1990s, call centers improved their automated understanding of callers by purchasing computer databases to store information about individual customers that could supplement what those customers told agents over the phone. These databases allowed organizations to maintain lists of customers' characteristics—from names and addresses, to history with the firm, to scores describing their value to the firm—that no human beings could possibly manage. A rush to use these tools led to a new term, "customer relationship management" (CRM).

The umbrella description for these developments, computer-telephony integration (CTI), describes the goal: to enable computer and telephone systems to interact. As the sophistication of databases increased in the 1990s and beyond, CTI supplied telephone representatives with information about customers that they had not previously had access to. Right from the start of the call, the agents could authenticate callers by comparing their phone numbers with the ones listed in the company's database. Screen popups and other tools gave the agents a dashboard profile of the customer and sometimes allowed the agents to include in their conversations an acknowledgment of a caller's history with and importance to the firm.

The rise of the commercial internet in the 1990s added to the torrent of personal data. Primarily to save money, call centers traded their traditional wireline methods for the internet's packet switching mechanism for phoning (a technique called voice-over-internet protocol, or VoIP). That allowed them to connect their widely separated call centers much more cheaply than in the past. But linking to the internet held another benefit: it allowed centers to capture not just what people said over the phone about a center's corporate client, but also what they looked at when they went to the client's website; what they wrote in emails, text messages, and chats to the firm; what they posted on the firm's Facebook page; and, by the 2010s, what they bought in the firm's online stores or on its mobile app.[15] As the twentieth century turned into the twenty-first, practitioners called this tracking an "omnichannel" approach that captured the "customer journey," and industry executives began saying that they were in the contact center, rather than call center, business. An executive involved in implementing these activities said in 2012 that "one of the greatest benefits is that now, because of VoIP, contact centers are able to more easily capture 100 percent of their inter-actions. This massive corpus of customer conversations is a very rich source for analytics."[16] Yet it raised a dilemma: having so much information about individual callers was great, but how could a human agent absorb it all during a phone interaction?

The question concerned more than the future of human versus technical resources. It held enormous implications for the future of those who would construct profiles of customers—humans versus AI-driven computers. Business pressures pointed to using artificial intelligence as much as possible: labor costs were rising, and the questions that callers were asking agents were growing more and more difficult to answer. A 2008 Contact Center Satisfaction Index report by the service-ranking company CFI Group confirmed that customers increasingly used calling firms "as the resource of last resort," turning to them only after they had failed to answer their own questions digitally. One consequence, according to the report, was that "in today's multichannel environment, customer service representatives are more likely to get a higher proportion of 'harder' questions that customers cannot find answers to on a Web site or elsewhere."[17] In that environment, according to CFI, one in five customers reported they could not resolve their problems with the contact center reps. That was an ominous sign, because CFI saw satisfaction with the contact center as an important indicator of loyalty and customer recommendations. The firm found that 94 percent of satisfied customers said they would do business with the same company again, and 91 percent would recommend it. Among dissatisfied customers, only 62 percent said they would remain customers, and only 39 percent would recommend the firm. "Customer service representatives are on the front lines of a company's interaction with their customers, so it's vitally important that they have the training and resources to do what customers expect of them," said CFI Group's CEO. "If customers just wanted to hear a friendly voice, they'd call their mom—but they are calling to get something done."[18]

Call industry executives, meanwhile, did not share the notion that contact center employees could be sophisticated and efficient handlers of torrents of difficult calls while also taking the customer journey, background, and relationship to the firm into account. The executives' more immediate concern was costs. Marketers, seeing the need for 800 numbers as well as opportunities for data capture, caused the call center industry to balloon in size, technical complexity, and competitiveness. Between 1988 and 1998, the number of U.S. companies involved in inbound or outbound operations (and often both) tripled to about 2,500. The difficulty of cultivating human talent at the wages the centers were willing to pay in such a fiercely competitive environment led them to adopt a strategy very different from the one advocated by CFI: paying agents as little as possible while ramping up the personalized information that technology could present to agents to satisfy callers. It became clear that while call-center leaders often hyped the rollout of ever more sophisticated customer management systems as efforts to know more about the customer, they took this step as part of a furious drive to lower the costs of speaking to the deluge of customers. As one executive noted, during the late 1980s and early 1990s, handling calls could cost a center more than twenty cents per minute—and with thousands of toll-free calls, that could add up. Live agents, especially U.S.-based agents, became a pain in the wallet. Consequently, "shortening the time on the phone by pre-populating data fields [with personal information about the caller and the caller's relationship with the firm] had a rapid ROI [return on investment]."[19]

Still, for many large marketers, U.S. call centers weren't bringing down costs enough. They began to use call centers in countries where wages were far lower than in the United States, a move made possible by the new internet phone systems.

Contact-center firms created internet-driven private branch exchange (iPBX) systems that moved incoming calls onto their VoIP corporate networks, converted the calls to compressed data, and routed them across the internet to wherever the agents were located—near or far, the cost was about the same.[20] Labor costs in countries like India and the Philippines could be as low as $1 per hour, compared with $6 to $10 an hour in the United States.

According to the U.S. Bureau of Labor Statistics, between 2006 and 2014, the United States lost more than 200,000 contact center positions.[21] Firms continued to push their costs down while installing technologies to quietly understand callers' backgrounds, respond to their demands, and guide discussions toward a conclusion that would make them happy customers. As one website for a firm selling CTI suggested in 2019 to harried executives, "Your team is handling more calls than they can manage. The phones won't stop ringing, and customers aren't being helped quickly enough. Stress builds for employees, which consequently gets felt by the customer. CTI can change that."[22] But supervisors were not about to let workers relax once the technology had helped to allocate the calls and give them information about callers. For while it enabled the caller to be more of an open book for the agent, the technology also made the agent an open book for supervisors. One website description of CTI's call monitoring and recording functions said they would "give management insight into how employees are performing and how customers are being helped. The monitoring function enables managers or coaches to listen in on the call and help guide the agent."[23]

The tensions around workload and offshoring spilled out into labor battles in the United States. In 2012, describing union organizing attempts at a call center in Asheville, North Carolina, a writer for the online Daily Kos called these centers "the sweatshops of the modern era."[24] A 2014 piece in Gizmodo struck a similar note: "The call center system as a whole is broken. And as you'll see from the tales below, it's breaking its employees along with it. . . . We've compiled some of the more appalling stories [sent by readers]; the recurring themes of debilitating stress, impossible standards, and wildly high turnover [rates] are too prevalent to ignore."[25] These domestic problems notwithstanding, the Communications Workers of America released a report in 2009 arguing that "the off-shoring of call center jobs is . . . bad for American workers and communities and harmful to the security of U.S. consumers' sensitive information." The report highlighted "a range of fraudulent and criminal activity emanating from overseas call centers," especially India, the Philippines, and Mexico, that included credit card theft, identity theft, and the illegal sale of customer data. Concluding that "U.S. companies have been exporting call center jobs by the thousands in a global race to the bottom," the report advocated passing "the bipartisan United States Call Center Worker and Consumer Protection Act." Sponsored by a Democratic congressperson from Texas and a Republican congressperson from West Virginia, this bill would have "required that U.S. callers be told the location of the call center to which they are speaking," that call centers offer callers "the opportunity to be connected to a U.S. based center," and that the U.S. Secretary of Labor create a public list of "bad actor" companies that offshore their call center jobs from the United States and make them "ineligible for certain grants and taxpayer-funded loans." The bill never made it out of any of the four committees that the House Speaker asked to consider it.[26]

The takeaway message of this dispute for the industry was that human labor, wherever it was located, would create trouble for contact centers and the companies they serve. Consequently, while consultants kept repeating the decade's mantra of cultivating customer satisfaction through omnichannel relationships—which often resulted in phone calls—the emphasis increasingly moved away from the ability of the human agent to the utility of the technology. In particular, industry practitioners increasingly relied on computers to create the personalized understanding and messaging that had historically been the task of people on the company end of the phone. When a trade-site editor asked an executive for the call center firm InfoCision to speak about the future, he mentioned not his call agents but technology. "At the heart of any CRM strategy," he said, "is the telephone channel, which provides a higher level of personalized communication. . . . The Internet gives customers more options to contact you—email, chat, social media, which have given way to increasingly higher expectations when it comes to customer service." That, he continued, "coupled with the struggling economy, has really pushed companies to new levels of efficiency—looking for new ways to produce ROI."[27]

It was no accident, then, that as early as 2005, AT&T created a voice assistant to help Panasonic field torrents of calls from customers about products they had bought. ("We were drowning in calls," recalled Panasonic's vice president of customer service.) The AT&T system identified key words among a caller's phrases and sentences and produced a reply in a female voice. It worked with simple problems that could be recognized through key words. When the system couldn't distinguish the words, it routed the call to a live representative. Basic as this was, Panasonic claimed the voice assistant lowered the average cost of resolving a customer issue by 50 percent. Success inspired imitation. US Airways, for example, introduced a phone assistant (this time with a male voice) explicitly to save money on human agents.[28] Both companies proudly noted that customers hardly seemed aware of the computer's presence, saying "thank you" as if they had talked with a real human.[29] In a further effort to gin up phone reps' productivity, contact centers began to use AI to discern the caller's mood. "Certain emotions are now routinely detected at many call centers, by recognizing specific words or phrases, or by detecting other attributes in conversations," wrote two *New York Times* technology reporters in 2010. They added that Voicesense, an Israeli producer of speech analysis software, had developed algorithms that it claimed could measure a dozen indicators, including breathing, conversation pace, and tone, to alert agents and supervisors when callers "have become upset or volatile."[30]

Many in the direct marketing business were coming to believe that humans—callers—need relationships, but not necessarily with living people. New developments in voice creation, voice recognition, and machine learning could lower labor costs while taking to new heights the ability to profile individuals and personalize messages for them. Nobody among contact industry leaders dared suggest that they would take human agents out of the equation. But Amazon, Google, Apple, and Microsoft were betting it could be done. First motivated by a desire for competitive advantage, then by a desire for customer surveillance, they would stress personality and personalization, knowing that these seductive features were the most likely to keep people engaged.

Their work would in turn accelerate contact centers' development and use of humanoid assistants.

Despite their importance to marketers, most of the early activities around voice in contact centers stayed below the public radar; the centers didn't inform callers about surveillance or what they were learning from it. Apple's Siri was the first assistant to interact openly with the public, and it created enormous enthusiasm for the potential usefulness of artificial intelligence in everyday life. Focused initially on speech recognition rather than biometric identification or inferences, this omnipresent character's benign affect eased the public into a marketing world where speech recognition and profiling for personalization would merge.

Siri did not start under a marketing umbrella. It was born out of taxpayer money in 2003, when DARPA funded the non-profit research institute SRI International to build a virtual assistant. Voice-activated controls and speech recognition features with various levels of ability had existed in home computers and other equipment starting in the 1990s, and DARPA hoped a more sophisticated interactional software would help military commanders deal with information overload. Called the Cognitive Assistant that Learns and Organizes (CALO), the project and its $150 million in government backing attracted hundreds of artificial intelligence experts. When they did develop intelligent assistant software, the successful result encouraged a number of business-minded engineers in 2007 to leave SRI, license key software from the CALO project, and develop their invention for the new iPhone. (A 1980 law made all that legal.) Reasoning that it would be a lot easier to use the Apple device through voice commands than by typing, they created an iPhone app called Siri (after the SRI mother ship) that was ready to go in February 2010. Steve Jobs had noticed; he may have seen Siri as a valuable rival to the Voice Search app that Google had recently introduced for the iPhone. Within weeks after Google's deployment of its app, Apple bought the Siri engineers' company, and over the next year it adapted Siri to its needs by reducing some capabilities (for example, its use of many outside web services to get information) and adding others (for example, several more languages). Apple also seems to have brought in Nuance to help with the backend technology for speech recognition.[31] When it released the iPhone 4S in October 2011, Siri was built in.

Although the CALO team members griped about the new owner's changes, Siri electrified the technology world. Google had announced the Voice Search app for its Chrome browser that June, and observers had recognized it as a breakthrough in voice recognition accuracy. Google had figured out how to recognize a person's voice request, transcribe it, and return relevant websites as if the person had typed the request. Yet as remarkable as that achievement was, Siri went well beyond it. Here was an entity on your phone you could ask to tell you facts or post a calendar appointment, and it would cheerfully do both. Articles commended it for its unique speech, crisp answers, and ability to joke, though the consensus was that the assistant wasn't as accurate as it should be (the Piper Jaffray investment bank and securities firm gave it a grade of "D" on understanding and answering queries).[32] Most observers gave Google's Assistant better marks for understanding.

Apple's success with the iPhone and Siri led, in a circuitous way, to Amazon's release of Echo and Alexa in 2014. It all started, ironically, with the enormous failure of Amazon's Fire Phone, a debacle that led the company to quickly pull the plug and announce a $170 million accounting loss. It was easy to understand why Amazon CEO Jeff Bezos would want to release a phone. In an increasingly mobile world, shoppers would buy more and more things on the move, with mobile devices. Amazon wasn't selling electronic (Kindle) books through other phones' app stores because doing so could mean having to give the hardware owner a cut of sales—in Apple's case, 30 percent. An Amazon phone would avoid those charges.[33] Perhaps more important, an Amazon phone would give the company real-time data about its customers and the ability to personalize its responses: Amazon would be able to track phone owners' locations, send them product recommendations based on that data, and use their whereabouts to build up their profiles. The challenge was to get people interested in such a product. Bezos thought the phone should include a number of unique abilities, such as a sophisticated display that looked like 3D on which the user could start apps by tilting the phone in different directions. As it turned out, those gizmos made the Fire Phone as expensive as an iPhone. Reviews were mixed, sales were terrible (analysts estimated only a few tens of thousands), and the company discontinued it in August 2015, barely a year after its debut.

But there was a silver lining to Amazon's failed experiment: the Fire Phone's development had involved work on a voice assistant. An executive in charge of the phone, Ian Freed, showed Bezos an early version of its software, which was able to recognize the utterance of any popular song title and then play it. Bezos was intrigued, and a few days later he asked Freed "to help build a cloud-based computer that responded to voice commands, like the one in *Star Trek*."[34] He gave the team a $50 million budget to hire speech scientists and artificial intelligence experts to create software that could recognize and respond to a far greater range of speech than song titles. Only four months after the calamitous Fire Phone release, the Echo, with a kernel from the ill-fated phone, made its debut. The team chose the name Alexa for the accompanying voice assistant out of a belief that while pleasant, it is unusual enough that users wouldn't often say it accidentally. The initial $100 price for Amazon's Prime members reflected the main takeaway from the Fire Phone flop: Amazon Senior Vice President of Devices David Limp believed that his division had priced the phone too high. Echo would be priced low, to draw larger audiences for its voice assistant.[35] An unstated consequence was that the profits from the device would come from other sources, including its surveillance activities—that is, from the company's savvy capitalization of Alexa's interactions with Echo owners.

As Amazon's smart speaker became a hit in late 2014 and early 2015, Google executives raced to match it. According to several accounts, Google strategists were not surprised that a virtual assistant would gain traction with the public, but they had been sure it would happen on smartphones and tablets.[36] Amazon had pivoted to the home only because it had failed with its phone. Google, rushing to catch up, released its Home smart speaker in the United States at a competitive price almost exactly two years after the Echo's debut, then pushed it out in more countries and languages than Amazon had been able to do. Apple's response was far slower. It began taking orders on its HomePod speaker with Siri in January 2018, a little more than three years after Amazon started selling the Echo. Clearly reaching for the high end of the market (much like the iPhone), the HomePod emphasized stellar sound at a price more than a hundred dollars higher than the original Echo.

Microsoft had joined the personal assistant fray in 2014 with Cortana, which it aimed to include in a future Windows operating system along with a Microsoft phone. And in 2017 Samsung introduced Bixby: linked mainly to smart TVs and appliances, it was the least used of the five in the United States. The common speculation of the marketers I interviewed was that each device had a different business model. Google and Apple had the largest numbers of people using their voice assistants—in the billions worldwide—because of the widespread use of the Android and iOS operating systems on phones, tablets, and computers. Google, with its legacy of selling marketers the ability to reach people through internet advertising, saw voice as the new way to search the internet—and a new way to track users doing it. As one analyst wrote, "if even a fraction of searches shift from mobile and desktop to voice interfaces, then that is where Google services need to be."[37] Apple, not committed to advertising as a moneymaker, positioned itself as a privacy-aware firm whose interconnected devices would work seamlessly. After a couple of years, Microsoft decided to drop out of the general voice-agent competition. Instead, leveraging its strength in business software and computing, it would position Cortana as primarily an assistant that people could use to plan their business day (via calendars, contacts, and email) and to help with business calls. Amazon's strategy was about selling products, its own and others, through Alexa. Some observers believed that Alexa's compatibility with Amazon's music, video, and audio-reading services was designed to encourage people to join Amazon's Prime buying program, which would lead to increased overall purchasing from the company. Others, not disagreeing, saw Amazon's goal more broadly. It was, in the words of one analyst, to "take a cut of all economic activity."[38] They saw sales via Alexa as another example of that.

Amazon and Google had the most interest in exploiting their assistants' surveillance features to sell things to users, whereas Microsoft and Apple wanted to know a lot about their customers in order to personalize their services to them. To accomplish either goal, each company needed to shape its assistants to keep current users interested while also attracting new customers. "Wired for speech" devices had to ingratiate themselves deeply with their users, and a key strategy was to imbue their voice assistants with personality. Strong humanoid-to-human connections that encouraged friendship and trust would mean fewer questions about the data their assistants were taking and using behind the scenes. Although the ingredients differed, each company followed the same basic recipe in concocting its voice character: First, imbue the voice assistant with a personality with which people want to engage. Second, give the assistant the ability to manage data about every user in ways that help those users get things done as successfully and seamlessly as possible. Third, place these assistants in devices that not only lure the user with what the industry calls "frictionless" benefits, but also allow the company to harvest voice and other data from that user across as many venues as it can.

The personality part of the recipe had precedent. Back in 1995, Clifford Nass and his colleagues at Stanford gave the creators of computer personalities a blueprint. They found that individuals could readily recognize personality types. Further, they said, the generation of a personality that moves people "does not require richly defined agents [or] sophisticated pictorial representations. . . . Rather, even the most superficial manipulations are sufficient to exhibit personality, with powerful effects."[39] The creators of the voice assistants, having intuited this from the start, based the robots' traits on bits and pieces of popular culture. Jeff Bezos's *Star Trek* reference in his instructions about what ended up as Alexa wasn't at all unusual. Read about the genesis of any intelligent assistant and you're likely to come across references to science-fiction and video-game characters. Martin Cooper of Motorola, who invented the cell phone, said the design was inspired by Captain Kirk's flip-top communicator on the original *Star Trek* TV series.[40] The fictional computer on the *Star Trek* ship, the USS *Enterprise,* also used a female voice to respond to crew members' requests. In the show, the voice belonged to the actress Majel Barrett, the wife of *Star Trek*'s creator, Gene Roddenberry, and while Google's engineers were developing their Assistant, they named it Majel.[41]

Cortana, Microsoft's phone assistant, is named after a twenty-sixth-century artificially intelligent character—"ever faithful companion of the Master Chief"—in the *Halo* video game series.[42] In fact, the actress who voiced Cortana in the video game, Jen Taylor, also contributed her voice for the U.S. version of the assistant. Higher-ups at Microsoft considered a different moniker for the public version, but a petition on a Windows phone user site evidently persuaded them to keep the *Halo* name.[43]

The 2011 version of Apple's Siri occasionally quoted Hal, the sentient and ultimately malicious computer in *2001: A Space Odyssey* (whom the American Film Institute named the thirteenth greatest villain in the history of movies). When Siri didn't know an answer, it would repeat Hal's well-known "I'm sorry, Dave, I'm afraid I can't do that." If a Siri user referenced a famous scene in *2001* by saying "Open the pod bay doors," the agent would reply, "We intelligent agents will never live that down, apparently." And in 2017, more than one observer saw Hal in the bright red circle that appeared and then swirled at the top of the new HomePod. "The glowing orb responds, when you're talking to it, just like HAL 9000," commented a Gizmodo writer.[44]

Marketers quickly decided that such inside jokes (dubbed "Easter eggs") should be turned into selling points. In 2016, Google enlisted Ryan Germick of its Doodle section (the group that creates the cartoons above the search box), along with Emma Coats, an animator who had worked for Disney's Pixar Studios, to give its newish Google Assistant "a little more personality," not only on smartphones, but also on the just-released Google Home speakers.[45] Google's CEO, Sundar Pichai, had said the Assistant was meant to be an "ambient experience that extends across devices"—for example, handing off between the phone and the speaker.[46] Immediately after she was hired, Coats set about giving Assistant a dramatic, tumultuous backstory so that users would empathize with it. Yet in 2017, with some experience behind her, she scaled down her ambitions to match the understanding of Nass and his colleagues. She described to *Wired* magazine how Google Assistant's "easygoing, friendly" personality was constructed by thinking up questions that humans are likely to ask Google and deciding on several responses for Assistant to use. Humor, she said, is good for both building the character's personality as "the fun, trusty sidekick" and for taking people's minds off mistakes or misunderstandings that might call attention to the character's non-human status.

"We don't want to have to fall back on something like, 'I don't understand,'" Coats explained. "That draws the attention back to you instead of continuing the conversation you're building." She also posited limits to the personality her team could give its creation. "Assistant can't be opinionated: it's there to be reliable, not to have depth." Nor could they write any script suggesting that the character is a tortured soul; "If we gave it some dark conflict secret, that probably wouldn't be a great user experience."[47]

Google wasn't alone in this project of creating a character that people could relate to from a modest script. Amazon writers tell of adding dozens of "delighters" to Alexa—including giving her groan-worthy dad jokes and concocting Easter eggs—typically inside jokes in response to certain questions or statements. Sometimes Alexa channeled old movies, like the comedy *Airplane*. (Human: "Alexa, surely you can't be serious." Alexa: "I am serious, and don't call me Shirley.") Heather Zorn, the Alexa team's director of customer experience and engagement, said the goal is to make the AI both useful and fun. She added that her team built their assistants' comments around several Alexa personality traits, trying to make her smart, approachable, humble, enthusiastic, helpful, and friendly. A Cornell University study that analyzed 587 customer reviews of the Amazon Echo showed that reviewers who referred to the device as "Alexa" and used the pronoun "her" were more satisfied than those who spoke of "Echo" and "it." Even so, Zorn asserted, the company doesn't want to turn Alexa into a member of the family. Instead, perhaps unwittingly echoing the Google engineers who created Majel, she said that her team's "guiding light" and original idea for the persona was the all-knowing ship's computer on *Star Trek*.[48]

Creating a persona perceived as a friendly and credible personality is the key to seductive bonding with a customer—to helping that customer feel psychologically tied to the device, and thus to the company.[49] Think first about the agent's voice. It became contentious in the United States that the default Siri, Google Assistant, and Amazon voices were all female. Critics argued that making a woman the default version of a polite, deferential, and pleasant assistant reinforced generations of harmful stereotypes of women in subordinate roles.[50] Samsung's Bixby stoked their anger as well, not because its assistant was female (the company gave people a choice of two genders), but because of what critics saw as "loaded, sexist" characterizations: in its language settings, Samsung described its female voice as "chipper" and "cheerful," and its male voice as "confident" and "assertive." (As a Twitter anger storm grew, Samsung quickly removed these labels.)[51] Yet neither Microsoft nor Amazon disputed that they were reflecting social stereotypes; both firms stated only that research with real people had led them to the gender they chose. In the words of a Microsoft executive, "For our objectives—building a helpful, supportive, trustworthy assistant—a female voice was the stronger choice."[52] Apple and Google seemed to feel that way too, at least for their American users, but they did roll out male voices for those who didn't want the default. In 2019 Google offered people ten choices—six female, four male—plus the "celebrity" pick of singer John Legend for some responses. Toward the end of that year, Amazon also moved a bit on gender, offering actor Samuel L. Jackson's humorously irascible male voice for a fee to replace Alexa on some activities. Unlike Google, though, it kept the female voice as the brand's enduring, easygoing persona.

To voice-first executives, ensuring that most users aren't put off was part of an essential seductive surveillance aim: making users feel comfortable enough to interact with the assistants and give them voice data and other information for profiling and personalized communication. In ads and instructions, the voice companies describe the ways an assistant can satisfy personal needs directly if owners allow them access to their voices and their lives: if given this access, an assistant can reliably post a calendar event, set a timer, answer a question via the web or Wikipedia, and much more, through the ease of speech. A 2019 Google Home video called "Hands-free help from Google Assistant" shows a kitchen on what appears to be a busy weekday morning. Two young adults are milling around, and a school-age child is eating; it's unclear how they are related. "The Google Assistant can distinguish your voice from others," begins the narration, "so when you ask for information on Google Home you'll get a response just for you. Get personalized briefings on your schedule, commute, weather, and more. So you're both ready to take on the day. . . . Call your personal contacts hands free." The two adults speak and the Assistant answers.

Young Adult One: Hey Google, tell me about my day.

Assistant: Good Morning, Alex. It will take you forty-five minutes to get to work.

Young Adult Two: Hey Google, tell me about my day.

Assistant: Good morning, Ross. Your first meeting is at 10 a.m.

The narration returns: "You can both request your personalized playlists using just your voice. . . . Open the Google Home

App to train the Google Assistant to recognize up to six voices. Once you're set up, everybody can start enjoying more personalized responses at home."[53]

The video reflects the view of Google CEO Sundar Pichai that the essence of the intelligent agent is the ability to personalize around every aspect of what a customer does and says. The responses that users hear from Google might seem to indicate the extent of what the assistants know about you and your life, but that impression is wrong. Unless you go out of your way to find ways to limit or delete specific types of information, the company reserves the right to "collect . . . voice and audio information when you use audio features," along with an enormous amount of other things it learns about the people who get their customized output.[54] This, to Pichai, is not only unproblematic, but also just the beginning. "Today we have an understanding of one billion entities: people, places, and things, and the relationships of them in the real world," he said at a Google Developer conference in 2016. "We can do things which we never thought we could do before. . . . We think of this as building each user their own individual Google."[55]

The CEOs of Amazon, Apple, and Microsoft could show similar videos from their firms, and they likely would agree with Pichai's conclusion. But in 2019, Google tried to show that it was ahead of the pack when it rolled out its Duplex technology. Available on iPhones and newer Android devices, it gave Assistant, with a male or female voice, the power to make reservations for the user. Observers noted that this iteration of Assistant sounded eerily human, even including halting sounds such as "uh" and "umm" (sounds called speech disfluency) in order to more perfectly mimic a person.[56] This naturalness caused concern during Duplex's initial public demonstration because Google didn't identify its caller as a robot; after the criticism, it began the calls with "This automated call will be recorded."

A person with a phone can simply activate the Assistant app and ask it to book a restaurant table, schedule a haircut, or check a business's operating hours using a choice of several male or female voices. When the reservation is made (within the next fifteen minutes), the Assistant sends the user a text message. If a call doesn't go through properly, or if the person on the other end doesn't want to be recorded, a human representing Google will take over. One reason for both the human involvement and the limited types of reservations possible so far is the difficulty of generating the huge training sets needed for Google's machine learning operations to figure out the best ways to discuss appointments. According to Google, the human operators who intervene are also there to write explanations of the glitches on the call transcripts used to train Duplex's algorithms.

Google positioned Duplex at the leading edge of personalization, a leap beyond customized phone voices and the logical next step in more personalized interactions. The voice was by now so personalized and friendly that Google was sure people would let it phone for them. It was all so seamless, at least in the demonstration, that questions about the surveillance that came with the seductive features—the information about their customers that Google and Amazon and others were taking and storing—didn't come up. The only indignation was about the impoliteness of a robot not saying who it was and that it was recording the conversation. A writer who had seen a demonstration made the obvious leap to Duplex's value for call centers. "Many big businesses are basically trying to make a human version of a robot," he wrote, by training them to "rigidly follow a script." The need for humans to mimic robotic assistants would virtually disappear if a large company got hold of "Google Duplex-style technology for its call center."[57]

That kind of all-purpose ability to converse with callers (or smart-speaker users) is not likely to happen for a while. As the Forrester consulting firm wrote, "The human brain has had millions of years to develop the architectural complexity required to comprehend and generate language."[58] Humans have linguistic abilities that computers still can't master. The big difference has to do with the uncertain meaning of many sentences and words outside of their larger context. A favorite example is the phrase "eats shoots and leaves." By inserting or removing a comma, heard vocally as a slight pause, you can make it about either a hunter or an animal. This sort of fuzziness is very much a part of people's everyday speech, and they can typically understand such phrases from the surrounding conversation. Computers, however, have a tough time making sense of these ambiguities, and this can lead to the frustrations that people sometimes feel with call center computers as well as with Alexa's generation of assistants.

Executives and engineers have taken aim at solving at least part of this problem, if only because they believe that training computers will in the long run be far less expensive than training humans. Nuance has for years been working with contact centers to use deep learning and other forms of artificial intelligence to improve computer interactions with humans. As early as 2013,

Nuance's chief technology officer acknowledged that "customer service transactions have proven difficult to organize into menu structures in a way that's efficient and understandable." In the future, he said, "specialized virtual assistants will provide direct access to information by bypassing the IVR [interactive voice response] entirely, and also will support flexible conversations that allow users to proactively provide unprompted information, and to jump freely among different contact center functions."[59] More recently, IBM, Amazon, Google, and Microsoft have entered the contact center world in competition with Nuance and others. Their pitch is that the well-known voice recognition and synthesis technologies they have developed in other business areas will work for contact centers as well.[60] IBM and Nuance in 2019 were the most aggressive in claiming that their AI technology would do away with the clunky IVR push-button choices. According to IBM, its Customer Care Virtual Agent, with the male voice of Watson, would allow customers to seamlessly speak to it "using the same conversational speech they would use with a human agent." Customers would no longer have to push buttons "in response to robotic-sounding prompts" or hope they had the right keywords to get the response they needed. Instead, they "can use complete sentences to interact with Watson, which can understand those sentences and select the appropriate, natural-sounding responses."[61] Nuance claimed that its assistant, Nina, will interact with callers as "a familiar voice [that] will answer their request whether it's typed into a computer, tapped on a screen or spoken into a device."[62]

Both have personalities. Watson is a bit aloof, but Nina can be funny. I asked it "Are you married?" and it replied, "Is that you,

Mom? Can we just focus on my job for now?" Their creators claim that each can personalize its conversation with an individual based on its ability to connect to the company's computers to learn the caller's background and previous interactions with the firm. And AI training is said to be easy. IBM contends that customization of "a basic conversation" can take as little as one to two days, and "within three to six months you should have Customer Care Voice Agent integrated into your call center system."

These claims make it sound as if marketers' future is here. Stepping away from its hype, though, IBM carefully acknowledges that its agent succeeds when people have routine concerns; those are the basis of the virtual agent's training set. Both IBM and Nuance emphasize that when people want to speak to a real person, their virtual agents "can quickly and easily transfer less routine cases to human agents."[63] The current state of the art is to give the human agent as much AI guidance as possible to efficiently deal with the caller's needs in ways that benefit the company. In a sense the goal is to create, with a combination of the human agent and an AI sidekick, an attractive persona that can be as tailored for that person in that situation as Alexa or Siri would be. Surveillance is central to accomplishing this goal. The firms carry out deep-learning analyses of the growing number of channels that individuals use as they consider purchases or look for solutions to difficulties they have with the products. As Forrester Research notes, tracing this customer journey means "combining quantitative and qualitative data to analyze customer behaviors and motivations across touchpoints and over time."[64]

Increasingly, too, analyzing the journey means exploring the customer's speech and voice patterns. Some in the voice intelligence industry are spreading the belief that even the most useful conclusions from a person's background and activities can be surpassed by deeper analytics that connect those characteristics to individual words and word patterns, and even the physical characteristics of people's voices. The goal clearly is to use seductive surveillance to help create an extreme version of personalization: to know a person better than they know themselves. And marketers are trying to access a torrent of speech, voice, and other new data to make this goal a reality.

2 WHAT MARKETERS SEE IN VOICE

Avi calls the 800 number of a web store where he is considering buying his wife a birthday present. He is thinking of a certain type of hiking boots, but he has never bought from that retailer and has some questions. He doesn't know that the store's call center uses intelligent predictive routing for upselling. As he dials in, the computer of the software company that provides the service matches his number with artificial-intelligence analyses of the words and grammatical style that he used when phoning other companies that the firm services. The computer concludes that Avi has a predominantly adviser personality—"dedicated, observant, and conscientious." It connects him to a human agent who, according to company research, ingratiates himself well with adviser types and is especially good at encouraging them to buy more expensive versions of the goods they are considering. It certainly doesn't always work, but in this case, using a time-worn script, the agent sells Avi the premium version of the hiking boots. The agent tries but fails to get Avi to buy another pair for himself.

Efrat enters an expensive dress boutique for the first time. Jackie, a salesperson, greets her and invites her to her desk to discuss her purchase interests before they look at specific clothes. Unknown to Efrat, there is an Amazon Echo inside Jackie's desk that starts recording a short portion of their conversation when Jackie says the wake phrase "make your day." Jackie's aim is to see if a specific promotion that Amazon rolled out some months before could apply to Efrat. Amazon had invited all of its adult

Echo owners to opt in to a service that gives points to people who allow specific merchants to identify them by their voices, and who then visit those merchants. The shoppers get even more points based on how much they spend on those visits; the more merchants and the more purchases, the more points per year. Someone who checked "all Amazon-approved merchants" and then visited a good number of them could earn enough points for a weekend vacation at a desirable resort. Jackie learns that Efrat has indeed signed up, and she is visiting the boutique because it is on Amazon's list. But because Efrat didn't read the privacy policy, she doesn't know that this dress boutique links the voice identification to information about Efrat's occupation, home location, estimated family income, and propensity to buy expensive products. This information comes in handy when Jackie chooses the items to show her customer. She politely ignores some of Efrat's comments about what she wants in favor of what the background information tells her Efrat can afford and what other customers from her tony neighborhood might buy. It works—or at least, Jackie believes that it does. Efrat leaves with two high-end dresses, oblivious to the exchange of data that allowed Jackie to try to manipulate her.

These two imaginary scenarios are a sobering indication of the claims the voice intelligence industry is making about its ability to personalize. The application that Jackie uses to identify her shopper can't be found in the wild yet, but the technology to implement it exists; Amazon received a patent for it in 2019 under the title "User Identification Using Voice Characteristics."[1] In Avi's case, that kind of upselling likely happens every day, and the intelligent predictive routing that was applied to his call is a product the voice analytics firm Mattersight has been selling since at least 2011. Both inventions simply profess to solve problems that beset people and companies in contemporary society.

Amazon's voice ID patent application positions its utility as erasing embarrassment. "For example," it reads, "if a user meets a stranger or a familiar individual whose name she may not remember, typically, the user may have to explicitly ask for the individual's name, which may be somewhat awkward in different social situations." Of course, marketers can use it for other purposes. Mattersight, which received a patent in 2014 for its "unique predictive behavioral routing capabilities," touts benefits such as "driving an immediate 10%–25% reduction in costs [and] improvement in sales."[2]

The claims for both technologies are rooted in a longstanding belief that the words people use and even the sounds of their voices provide information that they cannot filter and might not even know they are giving. Decode those words and sounds, the logic goes, and you can exploit them for many purposes, including personalized marketing. Analysts in companies and research universities believe that they are slowly finding patterns in individuals' physical voices and the words they choose that betray who they are, what they feel, and how they will act. As companies press on in the race to personalize the marketing of their products, voice analysis through intelligent agents promises an efficient way to learn people's personal characteristics. Voice analysis, supporters say, is often invisible to the target customer and so may not bother them, while for the company or group doing the targeting, it saves money and can't easily be fooled.

Voice intelligence researchers claim that they already can link your voice to your identity, body shape and age, social class, certain illnesses, and specific emotions and personalities. They also assert that they have been developing an understanding of the patterns in what you say as well as your sounds—even to the point of knowing the right voice and words to persuade you in any given place and time. The claims are audacious, many haven't been tested in the marketplace, and the technologies may not work as the researchers hope. Nevertheless, firms in the contact center business are already implementing some of these ideas. While they insist that their work reflects surefire understandings of individuals through words and voiceprints, their understandings are actually filtered through subjective decisions, definitions, and inevitable flaws in the system.

Google and Amazon, more in the public eye than their contact center counterparts, are far more wary about disclosing when and how they interrogate people's speech through their devices and search engines. The terms of service for their hugely popular voice assistants indicate that they're doing it, but they don't say how. An alternative way to measure Google and Amazon's intentions is to explore the patents they have won in attempts to control the future of voice intelligence technology. Those documents are all about ways to gain and use specific knowledge about individuals as the companies' devices hear them speak through the day.

Today, voice analysis is used mainly to prove who you are. The vehicle is voice authentication, which increasingly helps people who want to connect to their bank, investment company, phone company, or other account verify to a representative their relationship to that account. The authentication is part of a class of unique identifiers that includes fingerprints, facial recognition, and behavioral patterns (such as how a person types or uses a mouse) that contact centers increasingly employ to replace non-physical shibboleths as gateways into their domains. "PINs, passwords, and challenge questions are inherently insecure," states the voice identification firm Nuance on its website. "Fraudsters can buy credentials online and find personal details easily on social media." There's another reason: traditional passwords cost companies money, even with honest customers. The trouble starts with interactive voice response (IVR), the button-pushing routine that most call centers still use to route customers to answers (like what their balance is or where they should send a statement) without taking up a human agent's time. Nuance points out that "customers regularly fail their authentication" with PINS and passwords at the start of IVR because they don't remember them or push the buttons incorrectly. They then get transferred to an agent anyway, or after calling a number of times they learn to immediately press zero to talk to a real person. It can be a pain even with a human agent, though, to repeat identifying data such as name, address, birth date, and possibly a password. As a result, say voice authentication vendors, customers readily give up their physiological data to IVRs and agents. In fact, Nuance writes, "with multi-model biometrics, they can choose the authentication mode that matches their situation. On a crowded bus, it may be easier to authenticate with a selfie rather than a voice password. When customers are authenticated without having to be interrogated, both the customer and the agent are happier."[3]

There are various flavors of voice authentication, but the basics are the same. The first step is to use a digital recording to create a voice model that detects a person's unique speech characteristics. Those characteristics involve features of the person's vocal tract such as the length and the nature of his or her nasal passage (as reflected in the speaker's consonant and vowel sounds), as well as features like the rhythm, tone, and rate of the person's speech. Experts claim that these models are so robust they can distinguish between identical twins and recognize people if they have a sore throat or sniffle. Firms have different ways of creating the voice data. Some, like Google and Amazon, ask callers to go through a voice authentication procedure. A variation involves asking people to repeat a sentence such as "My voice is my password" up to five times. That creates the model and also a passcode they will use while their voice is analyzed. Quite often, though, the biometric initiation takes place without callers even knowing that the contact center is incorporating their voiceprint into a database. As an executive told me, "Your call may be recorded for quality purposes" is a cryptic statement that allows the firms to say they did their duty even for such voiceprinting and voice storage activities.

Once the model is created, it must be stored. Amazon places the voiceprint and text logs of the voice commands in its "cloud" servers, where they are used to personalize various applications for the individual and, for example, allow that person to complete a shopping transaction without the need to enter a four-digit pin.[4] Google, less clear on the subject, suggests that while it still sends voiceprints and text logs to the cloud, it has been moving to process much of the material on the device for speed and security, sending only "speech models" of the voice to the cloud, where it can conduct complex analyses.[5] Contact centers also may use cloud services, or they may choose local computer servers. The goal is to have an individual's voiceprint archived, analyzed, and ready to use as a reference model for when the person calls again. The contact center may assign different people's voice models values that indicate the model's quality—perhaps based on background noise—and its similarity to other voices in the system. When the person phones in, the computer uses the weighted model value against the caller's voice to decide whether it or the human agent should demand more information and, if so, what. Contact centers that serve many organizations may set up their servers so they can use the same person's voice data for different clients. They take this approach not only to validate honest customers, but also to keep fraudulent ones at bay. Having voiceprints of people who they believe have tried to commit fraud, whether successfully or not, gives them some protection against the next time, no matter what bank, investment firm, or other enterprise is the target.

On their websites and other public statements, firms claim that one's voice is the ultimate authenticator. "After the first seven to ten words, we have a pretty good idea that it's you talking to Alexa and to every internet-of-things device nowadays," Nuance's chief technology officer, Joe Petro, told me.[6] Yet a leading security expert told the United Kingdom's *Guardian* newspaper that "there is extraordinary technology now which is able to emulate people's voices pretty much in real time. If a criminal has fragments of you speaking already—for example, a YouTube video or podcast—there's technology that can put together a very convincing imitation of your voice."[7] In fact, though firms claim their systems can distinguish a person from their identical twin, in May 2017 a non-identical twin—a BBC reporter—got through HSBC's voice authentication system. Brett Barenek, Nuance's director of security strategy, acknowledged that fraudsters are becoming more sophisticated—for example, mimicking stress in order to fool the human or computer agent into letting them in. Still, he claimed, his company's software can discern "a difference between individuals under some form of stress and those going through the motions of some kind of stress." He added, "We've seen them try to use recordings. They try to emulate the customer's interactions." Barenek also said he expects to see malware that will attempt to record a person's voice and speech patterns. So far, he said, "they haven't been able to overcome our biometrics."[8] Yet he acknowledged that cyber-criminals continue to develop techniques for getting around security software and concluded, "We're definitely in an arms race." The collection of voiceprints, however, proceeds apace. Few leaders invite citizens to think about the implications of having a part of their body turned into a passcode that can never be changed but might be hacked by thieves or in the hands of government agencies. Despite the BBC break-in, HSBC said in 2018 that it was registering more than ten thousand people a month for voice authentication. Later that year, Nuance reported accumulating the voices of more than 300 million people for itself and clients. Commercial interests have made the giving over of one's voice feel like a normal part of modern life.[9]

In the face of its growing use, researchers and executives are convinced that authentication is merely the tip of the iceberg for what computers will help them understand about humans through voice analysis. Some researchers are academics who systematically explore voice with the help of artificial intelligence. Others turn out company patents that claim (without disclosing business models) to use AI to tailor consumer experiences based at least partly on what people's voices say about them. Someone with a foot in both camps is Rita Singh, a computer science professor at Carnegie Mellon University who investigates what can be learned about people from voice sounds. In a discussion about her work, she emphasizes that she's not interested in the languages people use or what they say. For her it's all about how the voice box reflects what's going on in other parts of the body as well as the mind.[10] Singh points out that what people's voices say about them has been a siren "to scientists, doctors, philosophers, priests, performers, soothsayers—in fact, to people of myriad vocations—for centuries."[11] During the twentieth century, research on voice was tethered to research about the physiology of speech and its relation to the rest of the person. Studies systematically "linked specific characteristics of voice and speech to the anatomy of the vocal tract, to the biomechanics of the voice production mechanism, to physiology, to neurological characteristics," and more. At the same time, quantitative diagnostic work explored links between voice features—sounds and their anatomy—and specific characteristics of human beings.

Singh surveys many of these links in a book filled with mathematical formulas. She details how factors ranging from skeletal to cellular features affect the ability to judge a person's gender through voice. (Males and females differ, for example, in the size and front angles of their thyroid cartilage.)[12] Certain emotions, too, affect voice production because of a relationship among the body's nerves. It turns out that "the laryngeal nerves that feed the muscles of the larynx have connections to the vagus nerve, among others." In turn, "the vagus nerve has been highly implicated in the body's response to emotion."[13] These sorts of explorations have yielded a catalog of other personal details that people unknowingly offer up about themselves when they talk. For example:

- Many studies have shown the relation between weight and voice; greater weight translates into a somewhat higher voice in men and lower voice in women. The changes relate to an influence of weight on certain dimensions of the vocal tract as well as (to a lesser extent) on different hormone levels. Researchers can even estimate the weights of five-year-olds by looking at the sound frequencies of their voices.[14]

- Statistical research has shown that certain characteristics of a voice can reveal the person's height within three inches. Taller people produce a different sound because they have, on average, longer vocal tracts and larger lungs.[15]

- Voice betrays a person's heart rate.[16] The reason probably has to do with effects of stress, such as adrenaline levels.

- Voice indicates whether a person is generally in good or bad health. People who are well have significantly higher vowel sounds, a wider range of sound-making ability, and lower levels of jitter in their voices.[17]

- Systematic deficiencies of voice are biomarkers of certain diseases. People who develop amyotrophic lateral sclerosis (ALS), also known as Lou Gehrig's disease, have voice deficiencies different from those with Parkinson's disease, which are in turn different from the deficiencies of people with dementia. Many endocrine disorders also have characteristic voice biomarkers.[18]

- The use of birth control pills is detectable in the voice within a few months after a woman has started to take them. Because the pills inhibit ovulation, they change the body's hormonal levels. Researchers can detect these changes in quantitative measurements of the voice's "range" and "quality."[19]

- Psychiatric illnesses such as depression, schizophrenia, and suicide risk have specific voice biomarkers.[20]

- Voice also reflects certain aspects of a person's environment. Researchers can determine, for example, whether a person is speaking while moving or stationary.[21]

Researchers in the twentieth century, when these findings were made, tried to dig into patterns of associations they saw between people's voiceprints and certain body features. The difficulty with that approach, Singh argues, is that it requires actually observing those relationships, but investigators generally cannot observe a subject long enough to detect a creeping illness or a subtle relationship between voice and emotion. The workaround? Artificial intelligence techniques, such as machine learning and deep neural networks, that let researchers discover patterns not observable to the human eye. The idea is to load the computers with voiceprints and—controlling for age, weight, and many other body features—let them find any consistent links between certain voiceprints and body characteristics of interest (such as illness, or the presence of caffeine or an emotion-based neurotransmitter). If the analysis finds such a relationship, the computer will be then able to test for it in individual voiceprints. Singh is optimistic that the voice will tell many tales. She and others have already made 3D re-creations of people's faces based on their voices. An MIT project, for example, used artificial intelligence to generate a rough digital image of someone's face based on only a short voice clip. As an article about the accomplishment noted, "Thankfully, AI doesn't (yet) know exactly what a specific individual looks like based on their voice alone. The neural network recognized certain markers in speech that pointed to gender, age and ethnicity, features shared by many people."[22] Note the word *yet*. It reflects the optimism, and a bit of the awe, regarding these new portals to our bodies and souls. Singh suggests that eventually it will be straightforward to identify from someone's voice a vast array of personal characteristics—perhaps facial appearance, body size, psychiatric illnesses, physical illnesses, age, intelligence, sexual orientation, drug use, emotion, and more.[23]

It's a stunning agenda, and although most of these techniques are still incubating in university computer science labs, psychology programs, and even business schools, powerful organizations outside the academic research community have been paying attention.[24] As word of her book's release circulated among computer scientists, Singh received inquiries from intelligence agencies.[25] Her work itself has been supported by the U.S. Coast Guard and the U.S. Department of Homeland Security, among other funders.[26] In addition to an interest in voice profiling for its military and intelligence uses, organizations are trying to develop AI-driven predictions derived from voice for a variety of business applications.

For her part, Singh believes that she and other scholars have only begun to demonstrate how the relationships between vocal sounds and human features can be analyzed and used. She is quite aware of how such information might be abused. It isn't hard to see how a voice analysis indicating psychiatric depression, the early stages of Parkinson's disease, or simply that a person is taking birth control pills can color how a person is seen by employers, potential employers, or insurance companies. Voice signals might encourage social typing by marketers, other organizations, and even individuals. Inferring your weight from your voice, for example, can be a boon to clothing retailers. And imagine a dating app that offers matching services based on AI scrutiny of a person's audio profile.

In conversation as well as in her writing, Singh worries that with the easy availability of machine learning packages, people with limited skills will be tempted to run analyses of people's voices, leading to so-called findings that are as dubious as the methods.[27] She also argues that some voice associations may be culturally based. For example, humans in particular social environments can agree on the emotion expressed when listening to the sound of a voice. But people from other cultures might not identify the same emotion from that voice. Stress is even more difficult. While it's not hard to note stress in a person's voice, linking voice characteristics to the cause of the stress is hard. (Notably, experts have concluded that the most hoped-for indicator from voice stress—that a person is lying—cannot so far be predictably detected by either humans or machines.)[28] When it comes to a person's voice reflecting his or her personality, Singh is especially cautious. Studies have shown, she admits, that people can consistently discriminate among a variety of personality traits through voice (for example, enthusiastic versus apathetic, proud versus humble, and sensitive versus insensitive). She neverthe-less insists that personality "is an extremely subjective character-istic" and therefore inappropriate for the "automated profiling" of voice that is at the core of her project.[29]

Singh's ideal choice for a voice profiling business, it appears, would be a technology firm that works with a health care organi-zation to draw conclusions about physical illness. She told the online magazine of a military security organization in late 2018 that she is a co-founder and "academic lead" of a startup, telling. ai, that will "initially focus on health care to aid early diagnosis of diseases such as Parkinson's and Alzheimer's and to more easily identify different kinds of intoxication, including from recre-ational drugs."[30] Although the company is not yet up and running, there is an older firm that links illness and voice profiling: Vocalis Health, an Israeli startup with an office in Massachusetts. Supported mostly by investor funding since its 2012 founding as Beyond Verbal, the company originally focused on creating AI-driven software to help clients decide the expressive state of individual customers according to a number of emotion groups and different levels of intensity.[31] The strong competition in that space, though, led the company to shift focus. It licensed its emotion technology to another firm and changed its primary research aim to using a person's voice to detect disease. Touting its training-set bona fides, Vocalis Health's website in July 2020 pointed to "over 1 million recordings of chronic patients" and "millions of data points" about them.[32] Vocalis Health added that "we are launching an initiative to collect your voices with a goal to be able to triage, screen, and monitor COVID-19 virus." Its more general aim, according to the website, is to use "proprietary voice technology" to create an "AI based platform that uses voice to detect and monitor health status."[33]

Early research indicates that the technology may work. A 2018 article by physicians at the Mayo Clinic who used Beyond Verbal's voiceprint recording technology did "suggest an association between voice characteristics and coronary artery disease." Moreover, the authors noted, other recent studies using methods similar to theirs have shown that "voice signal analysis can be used to identify autistic disorders with a success rate of 86%," differentiate "paroxysmal coughing from pertussis coughs" with 90 percent success, and diagnose Parkinson's disease with an accuracy of 91 percent.[34] While optimistic about these results, the Mayo Clinic authors do note that their small samples and other drawbacks mean more research is needed. Despite the academic caution, Vocalis Health's leaders are convinced they are right to focus on chronic physical illnesses. A senior executive (who wanted anonymity) said that chronic problems will be easier to detect and follow than acute ones, and moreover, physical illness is a better business bet than psychiatric problems. Other startups, he noted, are trying to use voice to diagnose depression.

This executive thinks the sweet spot with voice is to go after people who have chronic diseases or are worried about developing them. Vocalis Health would use phones or smart speakers to track changes in or the onset of specific illnesses via voice, with the goal of encouraging hospitals and health plans to integrate his firm's voice-tracking tech into their apps. That way, patients will use the voice technology with the blessing of health care providers and insurance firms, who will reimburse Vocalis Health. It would work this way: The app would regularly pick up several seconds of the individual's voice seamlessly through the microphone and send it to the cloud, where Vocalis Health software would analyze it for abnormalities. If none were found, no one would contact the patient. If issues did show up, someone from the health care provider would ask the individual to see a doctor for further testing. The senior executive said investors are enthusiastic about this vision. He hoped that a second article by the Mayo Clinic researchers, if accepted soon by a major journal as anticipated, would continue his firm's movement toward credibility. Yet he acknowledged that all companies in the voice-health space have a long journey before they can become profitable fixtures of the medical environment. A line on the company's web page reflects the slog ahead: "Vocalis Health's product & solutions might be subject to regulatory clearance and [are] currently not available for sale."[35]

While this slow-moving work has been taking place, a far more widespread use of AI to extract utility from voice has been unfolding. It involves drawing conclusions about a person's emotions and personality for marketing purposes. This effort has been complicated from the start because speakers' different cultural interpretations of emotions and the subjective nature of personality make it difficult to reliably label a voiceprint. Nevertheless, in their heated quest to sell new ways to discriminate among individuals, voice analytics firms have found it useful to claim that they can identify an individual's emotions and personality by using AI techniques on voice, alone or in combination with speech patterns. The firms arrive at their definitions differently, but they are united in the belief that they have discovered a way to capture exploitable signals that their corporate customers can use to understand and efficiently satisfy the people who call in, interact in chats, or send email.

You can see the competition most vividly among analytics firms that serve contact centers. In 2019, one such firm, Invoca, circulated a brochure called "Emotions Win" that reflects the general tenor of the contact business's approach. The brochure's subtitle, "What Customers Expect in the Age of AI," reflects the belief that in an era when so much can be understood about people through complex computer learning, consumers under stress expect companies to have "high EQ . . . the ability to recognize and respond to their emotional state." Invoca exhorts marketers to use artificial intelligence both to detect the emotions that individual customers are feeling, and to tell reps what to say in order to benefit from each customer's specific concerns. "A new frontier for customer experience is emerging," states the brochure. "For now, in-person and phone conversations are outpacing other channels in both sales and consumer preference for high-EQ interactions. Therefore, combining AI with human interactions is a massive opportunity right now and for years to come."[36]

Most contact companies are just beginning to grapple with the challenges and opportunities of using deep learning to draw conclusions about the people who call. From the contact center's standpoint, the trick is to do the analysis quickly, and to base it on what the individuals say in the moment in combination with other data the center or its client has gathered about them. Of central concern here is a tension reflected in Invoca's comment about combining human and AI interactions. Sales executives believe that many of their representatives would benefit from continual intensive coaching, but their companies "can't afford the human resources required to provide such advice."[37] They hope that artificial intelligence will step into the breach by giving agents advice, in real time, about what a caller's speech and voice really mean, and about what to do with that information.

It turns out that the conclusions that AI can draw about callers far outstrip anything the agent could learn to discern, and the technology is advancing rapidly. Just a few years ago the state of the art in caller analysis involved focusing on just a few characteristics of the caller and the call. For example, computers could prioritize certain callers over others based on their value to the firm as seen in purchases or (in the case of a bank or stock brokerage) money invested. Or a customer's mention of a competitor's product to an agent could trigger the computer to send the agent a script advising the agent how to respond. The computer could also predict from that customer's prior calls how likely he or she would be to respond to an upsell suggestion, and the system could tell the agent when and how to personalize the recommendation. In addition, by noting whether callers were mentioning advertising and promotional activities by the firm, executives could try to measure the effectiveness of their marketing campaigns and adjust or even personalize them based on the feedback.

Contact centers consider these activities useful. Yet the vendors that supply the industry with the software to analyze calls are thinking much, much bigger. For example, DialogTech touts that "When someone calls your business, the DialogTech platform tells you what marketing interaction drove that call, who the caller is, where they are calling from, and much more. . . . Contact centers can place callers from the marketing channels, ads, search keywords, or webpages with the highest purchasing intent into priority queues to get answered immediately. So if [an] insurance provider . . . thinks that Illinois callers from Google searches on cheap car insurance are better qualified leads than other callers, it can make sure that caller is answered right away, maximizing the chances they convert to a customer."[38] As early as 2017, a report from the Forrester consultancy firm called phone talk "an untapped gold mine." Customers, it said, present "a constant stream of high-quality data about who they are, what they want, and what they think about their experiences with your company." And yet the various ways they do it—comments to salespeople, give-and-take with chatbots, calls to contact centers, and far more—have traditionally been difficult to use for business insights, at least on any significant scale. In fact, "just analyzing what people say is difficult because of the nearly infinite ways in which people speak, let alone the unspoken context." The newest forms of artificial intelligence, Forrester asserted, would solve this problem. "AI, especially deep learning, is uniquely suited to handle the complexity of customer conversations—in terms of understanding both what is said and the context of the conversation." AI could help marketers discover "actionable insights" that were both new and previously unattainable.[39]

Analytics firms have jumped at the chance to exploit these latest opportunities presented by voice-related information. Nowadays, they all accentuate the need to use what the CallMiner call-analytics company calls "vast amounts of dark data" that contact centers collect from what people say—as well as when, how, and in what relation to other personal information they say it. "Dark data," CallMiner says on its website, "refers to information that organizations collect during their regular business activities, but do not currently use. . . . AI makes it practical to process and analyze this data" and use them as "a valuable resource to exploit."[40] Clearly referencing Forrester, CallMiner exhorted clients and potential clients to "unlock the value of your data. Speech is an untapped gold mine." In the same vein, NICE, one of the largest providers of contact center technology, stated that its AI-powered activities would "turn your customer interactions into valuable insights, and transform your contact center operations, your customers' experiences, and your organization as a whole."[41]

Behind the verbiage is a lot of technological elbow grease aimed at a few basic goals: call efficiency (typically meaning short handling time); resolution of the problem during the first contact; the caller's satisfaction with the call; the caller's respect for the center's client and desire to continue purchasing from it; more actual purchases from the company; the agent's ability to upsell the client; and the collection of information about the caller that will yield personalized results the next time she or he calls. The software firms have created AI programs to extract and analyze every communication a person has with the client via a website, Twitter, Instagram, Facebook, emails to the company, and more. The goal, in the words of DialogTech, is to "learn which channels, ads, keywords, and webpages drive the best sales calls, if new promotions are resonating, and why calls did or did not convert."[42] Yet phone conversations remain central to personalization. On its website, CallMiner emphasizes that it "leverages Artificial Intelligence . . . to analyze every customer interaction, across all channels, and automatically uncover actionable intelligence."[43] But a video on the same site notes that "more than 90% of customer conversations are still happening over the phone."[44] The language and tone of the phone call therefore remain central to CallMiner's—and its competitors'—efforts to draw conclusions about the people who contact their clients. On another video, CallMiner states that its Eureka cloud-based analytics solution "delivers an intuitive interface to measure intent and drive action. Expansive predefined conversation event patterning and machine learning enables instant insight from unstructured voice conversation."[45]

The call business analytics firms boast that by using artificial intelligence on the huge number of customer-agent conversations coming into their systems, they can predict the likelihood a person will recommend the firm to a friend or colleague (what the industry labels the Net Promoter Score), the customer's sense of how quickly and easily the company helps solve his or her problems (the Customer Effort Score), and a general score of Customer Satisfaction (C-SAT). Traditionally, companies have produced these measures through customer surveys, and many executives have been unsure of whether to trust those results given low response rates. Now, customers have no clue that a call center is turning their statements into pro or con viewpoints. One could argue that if callers knew their statements were being interrogated, they might alter their language to reflect a long view of the firm, not an opinion made in the heat of the moment, and others would be more emphatic about their concerns and dislikes. But the analytics firms contend that surreptitious drills into each caller's language provide insight into what callers actually believe.

Another part of this interaction that is not transparent is the contact center's use of data to identify and respond to each caller's emotional state, personality, and sentiment. The industry defines *sentiment* as a combination of attitude and emotion toward a specific company or its product. The speech analytics procedures of contact center computers begin with basic information about the interaction—the name of the customer, the day and time, the name of the agent. Text-to-speech software then turns the captured sounds into text. The system automatically explores the text for word and phrase patterns that training sets have associated with particular attitudes or personalities. In the same way, the computers can translate vocal sounds, and even silences, into meaningful units that industry practitioners call emotions.

With these procedures as starting points, contact-center computer systems draw many conclusions. They turn their interpretations of the language and emotions in any given phone conversation into a number that allows them to compare calls according to the quality of the agent's interaction with the customer and the customer's perceived satisfaction. They also combine their analyses of language and emotions into another number that indicates sentiment. Although many voice analytics firms claim that they detect "emotion" and "sentiment," they have their own secret-sauce algorithms for analyzing signals from customers and so interpret emotion and sentiment differently. In an unusually open display of an analytic firm's approach to interpretation, Verint stated in a 2018 report that it considers detecting emotion from words and phrases superior to detecting it from voice. "In over a decade of research and development," it claimed, "we've determined that the most precise mechanism to detect emotion is what people say, not how they say it. Pitch, tone, and other acoustic elements can give too many false positives when callers merely raise their voices or speak quickly to be heard, and miss complaints and dissatisfaction expressed while speaking normally." Verint asserts that it can confidently infer "different types of emotional reactions such as anger, annoyance, excitement, and confusion." Moreover, it can track when those emotions (as it defines them) show up in an individual's conversation with an agent, as well as sort those reactions into sentiments about the client or its products.[46]

Verint's competitors follow a different recipe, concocting an emotion from a combination of speech and voice signals. Their computers measure such characteristics as the speed at which the individual is speaking, the overall stress in the person's voice, and the changes in that stress as indicators of concern, frustration, or relief. Often the conclusions about emotions and sentiment are put to use after phone calls are finished, to identify what CallMiner dubs "the most impactful drivers of customer experience."[47] Increasingly, though, contact centers track sentiment and emotions in real time, so if a customer's mood is taking a negative turn, the computer or a supervisor can tell the agent to change the conversation in ways that the firm believes will increase satisfaction.

Neuraswitch is one of the firms that take personalization in that direction. Scott Eller, one of Neuraswitch's founders, told me that analysis of the words alone is not enough to determine callers' satisfaction with either the agent or the company the agent represents. A call might end, he said, with the customer saying, "Well, thanks a lot," and an evaluation of the transcribed text might report that everything went well. The computer's voice-emotion detector, though, may reveal that the customer spoke those words in an angry or sarcastic tone—a reaction that Eller says means callers are likely to soon abandon the company they were calling about. Neuraswitch therefore recommends that call managers have real-time access to customers' emotional reactions on all calls. The idea is to permit managers to intervene with the agent when a caller's emotional reaction is going downhill, when there may still be time to turn things around.[48]

Whether through speech alone or through speech and voice together, it is clear that capturing and analyzing customer interactions for automated scoring and sentiment analysis are the table stakes in today's contact center world. Firms are competing to interpret emotion and sentiment in what they claim are unique ways for unique insights. Verint highlights its ability to detect, in real time, a customer's risk of abandoning a retailer by tracking emotionally negative words in conversation with an agent. Its program then analyzes the caller's importance to the company and suggests what the agent should do. In a demonstration video, when a caller expresses anger and uses the word "frustration," a box with a caution sign pops up on the agent's screen that says, "Risk of customer churn. This is a platinum customer. You are authorized to offer a $200 credit."[49] The implication is that the system would not encourage making that offer to a less important customer.

Where Verint uses words to understand the customer's emotion, Cogito uses voice. A spinoff of MIT's Media Lab under the guidance of Professor Alex Pentland, Cogito encourages clients to "harness the power of voice" through "the world's only application that guides agent speaking behavior and automatically measures customer experience on every call."[50] The company performs "voice analysis" and gives the agent an "automatically generated experience score" that reveals the customers' supposed sentiment. It also provides real-time evaluation of a caller's "emotional states" (including the consistency of the caller's argument as well as his or her tone) and offers speaking cues to guide the agent's response based on how the system has categorized the caller.[51] The aim is to help agents adjust their speaking styles for "better outcomes."[52] The tech site ZDNet called this "artificial empathy."[53]

Then there is intelligent behavioral routing, an increasingly popular activity that proponents associate with a growth in customer satisfaction and revenues. Mattersight, in the early 2010s, was the first company to match a caller to an agent based on personality. Borrowing a proposition by the psychologist Taibi Kahler, the firm argues that "the way you and I communicate advertises our personality." It draws its notions of personality from algorithms that take account of the caller's words and word style. The firm's analyses look for six personality styles: connector (warm, sensitive, caring); organizer (logical, organized, responsible); adviser (dedicated, observant, conscientious); original (creative, playful, spontaneous); dreamer (calm, introspective, imaginative); and doer (adaptive, persuasive, charming). Mattersight executive Andy Traba explained to me that the vocabulary used to place people in one or another category is based on pattern recognition through machine learning.[54] He added that this categorizing happens after the phone conversation ends.

We first have to record the phone call in stereo because we want to be able to precisely identify the caller and the agent from that track. Then we transcribe that audio file into a text file, and then, through very sophisticated speech analysis, pick up on the patterns to make assumptions or assertions about individuals.

Those assertions can be the personality of the caller, their sentiment during the call (positive or negative), and even references to specific events—are they having a child, or getting married or moving to a new home? Maybe they inquired about a product they would like to be sold, or there was some degree of distress. Were they uncomfortable during that conversation?

Once you're able to identify or categorize all of those points, you can do the same on the agent's side to analyze how the agent's behavior or responses to those individual talking points influenced the conversation to a positive or a negative outcome.[55]

The company links that information to the caller's phone number and sends the result to a database tied to all its clients, not just the one whose product the person called about. Traba said Mattersight "has analyzed over a billion phone calls from our customers and we've taken those billion phone calls and aggregated them into a phone number database that has over 100 million numbers with associated personality style and the behavioral traits."[56] The clients never get the information (creating, the firm argues, no privacy breach), but Mattersight personalizes the caller's treatment through the agent on the assumption that the individual has the same personality across many companies.

Mattersight is owned by contact center powerhouse NICE, which has several Fortune 100 clients, so its intelligent behavioral routing categories have substantial influence. Other firms use different techniques to route calls to agents based on various definitions of personality and behavior. Neuraswitch, for example, draws its conclusions about a person's emotions entirely from voice rather than speech. It then links the voice-driven tags with data about people's behavior on the phone and off to create a personality profile. This leads to the process of "intelligently rout[ing] requests to agents that best match the personality profile of different customers."[57] It should be noted that not all of the firms that offer intelligent behavioral routing do it using speech or voice signals. Nevertheless, the analysis of what callers say and how they say it, in order to profile them and personalize future messages to them, is central to the contact center world. The application of artificial intelligence techniques to create a more extreme version of personalization based on speech-and-voice profiling is a natural extension of existing business practice. Mattersight's Andy Traba followed this reasoning into the future. "In the same way that there's information that's associated to an IP address," he said, "there's information that's associated to a phone number. We think there's going to be information that's associated with my voice print, which is as unique as my finger-print, and is going to allow these bots to look up a database then and say 'Oh, this voice print is Andy Traba; here is the personality and language he would like to be communicated with.'"[58]

The claim of every call-analytics company is that its AI-driven algorithms make those profiles accurate reflections of their customers' inner lives. The subjective nature of emotions and personalities at the heart of their work, though, suggests that to a large extent this is an expensive guessing game, disguised by jargon, that only sometimes pays off. The computational gymnastics generate practices that help some people more than others, for reasons the firms think they understand. As for the people who make the calls, the whole point of the process is that they will not know how to finesse the routines for their benefit—or even know that those discriminating personalization practices exist.

Then there are the newer endeavors for voice: smartphones and smart speakers. Understanding how and how much companies use an individual's voice data to make money via these devices is far more difficult than trying to figure out the same thing in the contact center business. Apple has a bigger stake than Google or Amazon in making profits off of its hardware, rather than through general retailing (Amazon) or advertising (Google and Amazon). Its privacy policies strongly imply that the Siri voice assistant simply comes with the phone or speaker products and that it makes no money on what it learns about you from your voice. Google, by contrast, admits that it does make money from your voice. Its privacy policy states, without going into details, that it can use any analysis of your "voice and audio information" as part of your profile for all sorts of money-making advertising activities.[59] The company carves out an exception for individuals who speak to its home-based devices such as the Google Home and Nest smart speakers, Nest thermostats, and Nest cameras. A separate notice, not part of the firm's privacy policy, assures these owners that Google will keep audio recordings "separate from advertising, and we won't use this data for personalization." It may, though, analyze the transcript of the audio recording—the text, not the voice itself—"to show you personalized ads." And if you speak to a voice app from another company (say, Spotify or Coach) that you have loaded onto your home speaker, Google will send a transcript of what you said to that firm. The "third party" firm has a right to then combine that speech record with other data it has gathered about you, take the information out of the app, and use it elsewhere.[60]

Amazon states that it allows firms using the voice apps on its smart speakers to do the same thing, but it is even less forthcoming than Google in disclosing how it uses your voice to make money. Buried in a densely packed FAQ is the news that "Alexa uses your voice recordings and other information, including from third-party services . . . to improve your experience and our services."[61] The company does note, in another densely packed document, "Alexa Terms of Use," that you can buy products through Amazon.com by talking to the speaker (the company can recognize your voice if you set up that feature), and that it may make "recommendations based on your requests"—for example, suggest you subscribe to Amazon Music if the tracks you ask for are not part of Alexa's music repertoire. The terms of use also state that Amazon helps those companies with voice apps on Echo devices to recognize you, and personalize their service to you, by giving them "a numeric identifier that allows it to distinguish you from other users in your household." You might also learn that "you can turn off automatic voice recognition, turn off personalization of third-party skills based on recognized voices, or delete any voice profile you've created."[62] But you have to sift through the extensive FAQ to find how to do that.

In other words, you can read through dozens of single-spaced pages of privacy policies, FAQs, and terms of service, and yet learn almost nothing specific about what the major voice assistant companies do with their customers' speech. But it *is* possible to get some insight on what firms think people's voices can tell them and how they might exploit that knowledge: look at the patents they turn out. Congress's ability to grant patents is built into the U.S. Constitution as a spur to innovation. A patent is a government license that gives a company or person the sole right to make, use, or sell an invention. The U.S. Patent and Trademark Office gives such licenses to technologies or devices its examiners have judged to be useful, novel, and non-obvious. A company can extend its patent coverage geographically by applying for patents in other countries through the Patent Cooperation Treaty, to which the United States belongs.

A patent attorney I interviewed, who wanted anonymity, pointed out that Amazon, Google, Apple, Microsoft, and other major tech firms are churning out large numbers of voice-related patent applications: "They are trying to cover everything under the sun, and they certainly have the financial capacity and manpower to do it." Many of these patents haven't been turned into commercial products even a few years after the Patent Office granted them. The reasons might vary, according to the patent attorney. "Sometimes you will file a patent on a device you are never going to make, just so no one else can. You can block the field doing that. Sometimes just the realities of life in manufacturing get so you conceptualize your idea [and] your R&D [research and development] team knows how to do it, but pushing it through to a manufacturing floor takes ten more years. So you can have your patent, but a product that is going to hit an end-user is nowhere in sight. So these big companies, they've got a team of in-house lawyers—all they do is meet with the R&D guys and see what they've come up with that month and crank patents out."[63]

Jeremy Carney, an intellectual property lawyer who is involved with voice technology patents, claimed that most companies do intend to use the patents they submit.[64] Yet he noted that even when products do flow from voice patents, they only vaguely resemble the descriptions in the patent application. Part of the reason may be that by the time the patent gets approved (often two to three years after filing), company tactics have changed and the inventions are adapted to meet new needs. But perhaps the more important reasons for the mismatch between application and product relate to the strategies and requirements of the patent process. The patent lawyers who prepare the application want to cloud its specific purpose from competitors. And as both Carney and Tom Cowan, another patent expert, explain, attorneys must prove to the patent examiners that an invention is novel and therefore distinct from previous creations (called "prior art"). They do that by describing some elements of the invention in such specific terms that an examiner can never find such unique elements in a prior application. But they also must craft their application broadly enough that future judges will find that a product having any resemblance to their invention infringes on their patent. The result, says Cowan, is that patent attorneys become experts at crafting their patent application to balance these two goals.[65]

Although this balancing act can make it difficult to discern from a patent application the inventions' actual intended uses, both Cowan and Carney agreed that the general trend of patents at any given time can suggest the industry's strategic direction and what developments are likely coming. In the case of voice technology, a review of the past few years' patents granted to Google and Amazon depicts a world where people freely give their voice data and other biometric information wherever they are in exchange for, as the companies portray it, frictionless and personalized service without negative consequences.

Let's imagine how the two firms' patented dreams for cutting edge personalization might be woven into a day in your future (fictional) life . . .

It's not just a regular day. It's a Saturday morning and your birthday. The Echo device next to your bed wakes you at 8 a.m., which it knows to do based on the calendar you coordinated with it. As usual, when you ask it for the news, Echo knows you by your voice, because right after you bought it you spoke to it through the app and said who you are. The device has long since determined how old you are. Patent 10,096,319, granted to Amazon in 2018, says that Echo's voice processing servers have the capability to "determine the user's age or gender" as well as whether you speak with a non-American accent, and in some cases (for example, Chinese) what your native language is. The technology can also "process the voice data to determine whether the user is in a normal or an abnormal emotional state"—and if it is abnormal, suggest something that might change your disposition.[66] That seems to have happened to you a couple of days ago. You felt a bit low when you asked Alexa to tell you the weather and read you the news; you were thinking of a plane crash that had happened the previous night. The Echo, in response, played upbeat rock music and (because you're an Amazon Music subscriber) asked if you wanted the song added to your Amazon music favorites. Today, though, Echo is quiet after the news. You say to yourself, "I must be feeling my normal self today."

You told your sister Jill, who lives around the corner, that you would eat breakfast with her and her family this morning. Two weeks ago, Jill and her computer-engineer husband, Jack, invested in a connected home that can help them entertain and discipline their two children, ages five and ten. Consequently, the home's cloud computer is learning patterns of the household via special microphones placed throughout the home that suggest who is where, in real time. Based on the pitch of voice signatures, patented Google circuitry infers age information—for example, one adult male and one female child are in a room.[67] Over time, the system's "household policy manager" will be able to discern such things as when people are at the dining table, when they are in the living room, how long the children watch television, and when electronic game devices are working. Jack's goal is to increase the amount of time the family spends eating together and lower the kids' time with TV and other electronics. As "policy administrator," he wants the system to report how much time the family members spend on all of these activities over a month. Then he will tell the system to regulate the TV and inform him if one of the children exceeds the game limits.

You worry that your visits will throw things out of whack—shorten their eating time and increase their game time because the children want to play with you on the PS6. But today is Saturday and your birthday, so you're hoping that Jill, Jack, and the AI system won't mind. You've done some reading on the approach Jack is pursuing, and it turns out what he's doing is just the tip of the Google-enabled iceberg. Other parents are having their homes monitor bullying through a combination of microphones and cameras that detect "audio signatures indicating 'bully' keywords such as derogatory name-calling, elevated voices, etc." And, says Google in patentese, the connected home has the ability to sense people's emotions:

> For example, a generalized inference of a happy emotional state of the occupant may be made when a visual indication of laughter and/or an audio indication of laughter is obtained. In contrast, particularly tailored inferences may look at the emotional context data in view of a particular occupant's known characteristics. For example, the system may know that Sue cries both when she is happy and when she is sad. Accordingly, the system may discern that an inference of Sue's emotional state based upon crying alone would be weak. However, the system may also know that Sue typically smiles when she is happy and maintains a straight face when she is sad. Accordingly, when a visual indication shows that Sue is crying and has a straight face, the system may infer that Sue is sad. This inference may be different for other occupants, because the particularly tailored example uses particular emotional indicators of Sue.[68]

Google's use of its dispassionate, seamless AI interface can sometimes be poignant: when there is a shared child custody agreement, the estranged parents can transfer the technology along with the kids, ensuring that the children will follow the same policies in both homes. This is not a concern with Jill and Jack, who seem to get along. But if they did split up, you suspect that Jack, as a nerdy and strict dad, would implement the shared custody technique.

After pleasant and uneventful breakfast at Jill and Jack's house (bagels, cream cheese, and smoked salmon, because no number of cloud-accessible recipes will make either one enjoy cooking), your voice profile follows you through the rest of the day. At a coffee shop that afternoon, a Google-patented app turns your tablet screen dark when it hears an unfamiliar voice approach too close, to keep your data private. When, on a whim, you decide to go to a store to buy something nice to wear out to a birthday dinner with your friend Diego, your voice profile guides the decision. The salesperson's app records your unconscious verbal reactions while trying things on, because these indicate what you like and don't like better than your words. The Amazon patent for the app notes that audio data can "indicate a user's mood," with the rate and tone of the person's speech assessed along with the actual words, to ensure accuracy. ("For example, the phrase 'I feel great' could have a positive meaning if said with great emphasis and at a higher rate—'I feel great!'—than if said lethargically.")[69] Following Amazon's suggestion, the salesperson starts off by bringing out clothes similar to those that elicited your most favorable verbal cues the last time you shopped there. You pick a blue blazer from that first batch. You have five similar blazers at home, but you figure the voice analysis must know what it's doing.

At the restaurant, you find your friend Diego waiting for you, looking annoyed. Over the past couple of weeks he has realized that when he asks his phone to read material to him on the subway, Google has somehow determined that he has low proficiency in English, and it sometimes changes the words to make them simpler. He is worried that some of the third parties whose material he is reading, and who may receive data from Google, may now think he is less qualified for work or education than he is. You wonder if algorithmic bias may be involved; Diego has lived in the United States for most of his life, but he has a slight Mexican accent. After you both order, you look on Google's website and find that a person's language proficiency score can be negatively affected by background noise when the person speaks into the phone. (Diego works for a contractor and sometimes makes calls from a noisy construction site.) It also notes that people can opt out of having personal information such as "information about a user's social network, social actions or activities, profession" tied to their score.[70] You promise to help him figure out how to do that in the next day or so.

Diego then tells you he had an odd situation with voice authentication before coming to the restaurant. Something went wrong with his Amazon Echo, and when he called customer service, the agent asked if it was okay to temporarily make the device believe that it belonged to the agent so the agent could correct the problem. "I trust Amazon," Diego says, "and I know I was talking to an Amazon rep because I phoned them. At least I hope I was talking to an Amazon rep," he adds with a nervous laugh. To make him feel better, you assure him it was an Amazon rep, but you wonder as well. It seems like Diego has been having voice-tech issues today.

It's getting late, past 10 p.m. You and Diego go over to your apartment for some dessert. Your alarm system recognizes your voice and lets you in. Immediately, an app on your phone that recognized your entry alerts you that "a teenage child . . . is out past his or her curfew," just as a Google patent predicted it might. Google's invention has allowed the neighbors in the area with

Google Home and Nest products to create a network that will alert authenticated individuals that wayward teens need to be caught, or, "in the event an occupant of a particular home contracts the flu or some other communicable illness," will send a notice to the authenticated individuals in other homes "so that precautionary measures can be taken to help prevent the young children and elderly adults from contracting the illness."[71] You groan at the alert ("they're chasing teens again," you think). But Diego, intrigued by the situation and remembering the coronavirus, says "that's a cool technology."

You both agree not to help chase down the errant teenager. Instead, you continue to celebrate your birthday with the most hilarious TV you've watched during the past year, using TV software patented by Google.[72] Over the past year you have allowed Google to collect snippets of your viewing and tag them according to certain emotions its computers say you've displayed through your facial and voice expressions, as seen by a camera and microphone built into the television. You and Diego can now ask Google to play back the snippets that elicited your laughter. It proceeds to play twenty five-minute segments, along with photos of your reactions while watching them. You and Diego find this a strange experience: for most of the clips, you can't quite understand what the big deal was. From the photos, it appears that Google thought you were in a hilarious mood when you don't think you were.

Still, you had fun watching the videos, and when Diego leaves, shortly before midnight, you delve into one more personalized feature of voice analysis. You turn on your bedroom light with your voice, ask Alexa to play your "classical music," and pick up the historical novel you've been reading. As you do that, the Echo Show next to your bed glows with a message about historical fiction that you might enjoy reading on your Kindle. Through "presence event notification," Amazon has determined that turning on your bedroom light at that time of day and then asking for classical music means you're reading a book, likely a historical novel.[73] Typically you read on your Kindle, so the Amazon computers have generated a reminder to coax you back to that habit. You sigh at the predictability of this activity. You used to wonder, as you started to fall asleep, what adventures you would face tomorrow. But that was when all these technologies were new. Now, as you doze off thinking about the reminder, Diego's problems, and the tailored videos, you marvel at all the ways that various companies categorize you via your voice, and how accepting you've become. "They are certainly trying hard to know us," you say to yourself before you fall asleep.

While none of these incidents was real, all were concocted from small suggestions in dense, dry documents. Amazon and Google are only two of the many firms patenting processes that depend on voice analysis, ultimately creating a universe where individuals give up their voice data without thinking about it. The case of Diego's language score is unusual in that the patent application itself warns that the innovation might cause harm and suggests correctives. But benign as the other incidents above may sound, they too reflect technologies that the patent owners could use for purposes that might anger the individuals involved. As comfortable as it might sometimes make you feel, you may not want people to be able to surreptitiously identify you via a voice app.

Having your semiconscious grunts and ahs be part of the criteria when you try clothes on may make some sense—but the notion of the merchant saving them as indicators of your personality may make you a bit queasy. And you might feel that having a company label you as being in or out of an "abnormal" emotional state is fine for suggesting music but for little else. A state tagged as "abnormal" could spell trouble if someone other than you gets hold of it.

You may feel that such great technology is worth the sacrifice, or you might not. Google and Amazon executives are motivated to help you think that it is indeed worth it; they want you to invite assistants to know your voice and respond to it, and to tamp down any of your queasiness about the unsettling parts. The incentives for this are financial, but in ways that you may not expect. For although some sources contend that Google is making strong profits from its smart speakers, others aren't so sure because of the huge discounts it often offers.[74] Dave Limp, Amazon's hardware chief, acknowledged in late 2019 that his firm wasn't generating much money from the sale of each device.[75] The companies are also not bringing in substantial advertising dollars from the smart devices because they are severely limiting ads. Instead, they are at this point encouraging a system that habituates Americans into giving up their voices to devices in various environments and situations. Limp said his goal is to make money when "customers use the products, not just when they buy them."[76] The overarching logic is that once masses of Americans feel comfortable with smart speakers permeating society, they will use them frequently and profits will follow. Everything voice-related—personalized advertising, web and app purchases, voice-activated product purchases, subscriptions for voice-linked services—will seem natural, friction-free, and difficult to refuse. Amazon and Google compete strongly with one another when it comes to selling items that respond to voice, but they have remarkably similar strategies for playing up their devices' seductive aspects and playing down the surveillance. Both are blanketing the nation with voice-first gadgets through public-relations campaigns and rock-bottom pricing. And both are allying themselves with organizations that place those gadgets in a multitude of places—kitchens, bedrooms, hotels, cars, schools, stores, and more—with the goal of helping huge populations become comfortable using voice assistants virtually everywhere they go.[77]

3 AN OPERATING SYSTEM FOR YOUR LIFE

"Amazon wants Alexa to be the operating system for your life," read the headline of a September 2018 article by Nick Statt, posted on tech site The Verge. "Amazon wants Alexa everywhere," the subhead added, "and it will even go head to head with microwave and subwoofer makers to get there."[1] In the piece, Statt confronted Amazon's recent "rapid-fire announcements" of Alexa-oriented hardware. A week earlier the company had held what one of Statt's colleagues called a "sprawling event" that showcased an explosion of Alexa-linked devices four years after the release of the original standalone Echo speaker.[2] The company was rapidly expanding its footprint in and out of the home. Among the products Amazon unveiled were

- an improved Echo Dot (a hockey puck–shaped version of the Echo) with better sound than its predecessor;

- an Echo Sub—subwoofer, that is, for giving sound depth to Echo speakers;

- an Echo Link, a receiver that puts Alexa on stereo equipment not natively linked to it;

- an Echo Input for giving Alexa's features to speakers not natively linked to it;

- a Fire TV recast, which allows video streaming and recording with the help of Alexa voice commands;

- an Amazon Guard feature on the Echo that can interpret the sounds of breaking glass or carbon monoxide alarms and send an email to the owner;

- an AmazonBasics Microwave, which will follow voice instructions via Alexa;

- an Echo Wall Clock, through which you can set timers, alarms, and change the time using Alexa commands;

- an Amazon smart plug, which allows you to use Alexa to start or stop anything that is connected to it;

- a "second generation" Echo Plus that can act as a hub for voice control of a range of smart home gadgets, such as lights, cameras, and thermostats;

- a "second generation" Echo Show, with a larger screen than the original, a better speaker, and (like the Echo Plus) the ability to act as a voice-control hub;

- a Ring Stick Up Cam, a camera for outdoor and indoor use that responds to Alexa requests to show the device's video feed on an Echo show. As the Ring website notes, "By saying 'Alexa, show my front door,' you'll instantly get a live video feed of activity at your home on an Echo Show"[3];

- and Echo Auto, designed for bringing Alexa into the car.

To Statt, this parade of voice-activated merchandise demonstrated a broad strategy of seduction and habituation. "It's clear now," he wrote, "that the company has every intention to make Alexa, its Echo line, and every single device open to integrating its digital assistant, into the dominating force in the home."[4] Forrester analyst Jennifer Wise agreed. "While today it looks like scattershot products being launched," she said, "there is a more measured long game here. . . . Consider these products to be a gateway into the customers' lives—one that down the road will make Amazon's intelligent assistant the one of choice over other options because Amazon has become so ingrained."[5] Statt added that Amazon would use the devices "as windows into our purchase behavior and as opportunities to steer that behavior in a direction that benefits Amazon." He suggested that the seductive additions to the product line—which emphasize personalized voice-driven results (sometimes on screens) and low prices—would lead to widespread habituation: "the more the company can lock you into its ecosystem, the harder it will be to leave it." Two Amazon employees told me that the more often Echo owners use their speakers, the more likely they are to purchase products through them as well as directly from the Amazon website and app. Writers in the trade press have suggested that Google would find the same pattern. It makes sense, then, that both companies would aim to flood the market with speakers and then cultivate habits among users that make giving up one's voice to marketers seem simple, natural, and helpful.[6] Wise pointed out that one tactic for pushing the Alexa habit is to link the voice assistant to all sorts of product categories—even seemingly odd ones such as microwave ovens. Amazon entices people into the Alexa routine by constantly experimenting with the device to see what sells and what doesn't, as well as by fiddling with the price, often keeping the cost of some Echo devices "ridiculously low."[7] The same could be said of Google and its smart speakers.

As Statt and Wise suggest, Amazon and Google are trying to create a new world where voice-first gadgets are part of the everyday landscape. So while the two compete fiercely over customers, they are working in concert toward three common social goals: seducing Americans into buying voice-first devices, playing down concerns about surveillance by these devices, and habituating Americans into giving up their voices every day and everywhere. Sometimes Google and Amazon carry out these activities by themselves, sometimes they do it with other companies in the voice intelligence industry, and sometimes other companies take the initiative. The entire industry is promoting the spread of voice-first technologies throughout key areas of society and encouraging the habituation that comes with it.

For many Americans, Amazon Prime Day is when their smart-speaker habits begin—sometimes, ironically, with a Google device. A bargain-hunting holiday inspired by Black Friday, Prime Day began in July 2015 as a celebration of the twentieth anniversary of Amazon's founding; by 2019 it was being celebrated in eighteen countries, including the United States and China. It's named after the Amazon Prime membership deal, started in 2005, that offers expedited shipping and other services (including Amazon Video and some parts of Amazon Music) for a yearly fee—$119 in 2019. The Cowan research firm estimated that in mid-2019, 53 percent of U.S. households had a Prime membership, which translates to about 60 million people.[8] The sales extravaganza of Prime Day actually now lasts for more than one day; in 2019, the discounting continued for forty-eight hours. Anyone can get the discounts simply by joining as a Prime member, but Amazon clearly sees Prime Day as the ultimate loyalty play: existing customers account for 60 percent of its U.S. sales. The day is therefore both a reward and reinforcement for being a Prime member and an incentive for people to join. Amazon surely knows that one way to get around the restriction is to sign up for a free thirty-day membership trial, and then cancel. The company nevertheless boasted that during the first day of Prime Day 2019, more new Prime members signed up than on any previous Prime Day. The selling power of those days is truly awesome. During the three-day Prime event of 2018, Amazon accounted for 86 percent of all internet transactions among the top forty-eight retailers.[9] After the 2019 event, which continued for four days, the company announced that it had beaten the previous year's take, selling more than 175 million items.[10]

Amazon has especially used Prime Day to populate homes in the United States and elsewhere with Alexa voice products. It dropped the prices on many of them substantially, leading to a rush of purchases. One website that tracks discounts concluded that in the United States, Prime Day 2019 was "defined by the best ever pricing on key Amazon products."[11] Echo smart speaker prices dropped 50 percent—the Echo and Echo Show (with their hubs) went for $49.99, the Echo Dot for $22, and a Fire TV stick for $14.99.[12]

The discounting worked. A July 2019 Amazon press release coyly titled "Alexa, How Was Prime Day?" announced that "Prime Day was . . . the biggest event ever for Amazon devices, when comparing two-day periods—top-selling deals worldwide were Echo Dot, Fire TV Stick with Alexa Voice Remote, and Fire TV Stick 4K with Alexa Voice Remote."[13] In addition, Amazon celebrated great deals on "connected home" devices—technologies such as Ring security cameras (Amazon owns Ring) and Ecobee smart thermostats (Amazon is the major investor in the firm), both of which can be controlled by Alexa.[14]

The success of Prime Day led Google to strike back with its own version of the sales circus. Walmart, Target, Best Buy, and eBay paralleled the Prime Day calendar with their own attractive deals on Google products. (Perhaps not surprisingly, Amazon didn't sell Google's smart speakers, though it did offer the company's Nest thermostat and camera products as well as smart plugs for Google's voice-first devices.) To counter Amazon's deep discounts, Google made a deal with Walmart to promote its smart speaker line at prices similar to Amazon's, while Best Buy and Target themselves joined the discounting fray. Walmart, Bed Bath & Beyond, and Best Buy offered a $150 discount on Google's Home Max smart speaker ($249 instead of $399). The stores priced their Google Home and Google Home Mini (competitor to the Echo Dot) devices to track the discounts that Amazon was offering on parallel products. Walmart also slashed prices on voice-activated smart security and assistant-powered products from Google and its Nest subsidiary—cameras, thermostats, and smoke detectors.

The game of hyping the seduction while minimizing the surveillance reverberates well beyond Prime Day. In fact, says Brad Russell, the integration of Alexa and Google Assistant into the home has three levels.[15] Russell is research director for connected home at Parks Associates, a marketing research firm that focuses on technology in the home. He says that purchases of smart speakers, smart thermostats, and smart cameras through carnivals such as Prime Day represent the lowest level of home adoption. The level is basic, he says, because these are "kind of do-it-yourself, standalone devices." This is the level at which shoppers decide whether "they're going to be an Amazon household or a Google household," but they don't apply that thinking to other devices or, especially, to a system of devices. They also rarely think of connecting future major appliances such as refrigerators and stoves with Amazon or Google assistants. Instead, he said, they hand-select devices for specific jobs. "They're hiring a device to do a job, and so they're looking for the best device in any category to do that job." People fall in love with devices because of the product's design, its features and the way it functions, its perceived reliability, and user support. They may not care if the products interact with one another, even as they enjoy being able to talk to their devices via Alexa, Google Assistant (many devices can connect to both), or less frequently, Siri.

Yet there are people who want, or can be persuaded to want, the joy of using voice to bring together the range of devices in their homes. This second level of adoption is where Nationwide Marketing Group (NMG) enters the picture.[16] Tom Hickman, the group's president, says NMG uses the membership clout of its 5,500 small- and medium-sized retailers to compete against big chain stores like Best Buy, Home Depot, and PC Richards. Its current strategy centers on helping stores help sophisticated do-it-yourselfers or curious people needing advice to turn their dwellings into personalized, connected homes. "The magic really happens . . . when all those smart things are connected together and solving real world problems" such as answering the doorbell while you're stuck in the kitchen cooking dinner. "Let me give you one kind of real life example from my own home," he says, when I ask him to explain why people should want voice-first devices.

So if my wife's got her hands in flour because she's making dinner and someone rings my front doorbell, my Nest doorbell cam can actually show up on the Google Hub that's on the countertop teaching her how to make that cake, and she could see and speak to that person and she can just tell them "Hey, just drop the package at the front door" or "Hey, come around to the side." That's a practical, connected application. You're actually talking about [two] different devices.

Another answer is—When you start to think about the future of connected devices—you're thawing this beef, say. Let's just say your refrigerator [has] all the sensors in there to find out when it reaches a certain temperature and then once it reaches a certain temperature, the stove comes on and starts to preheat so that you have a seamless transition in that experience.

These are real-life practical examples of how the inter-connectedness of all this really does make . . . life a little easier. [Consider] that you talk to your voice device and say "based on what's in my refrigerator right now can I cook a quiche in thirty minutes that feeds six? And your refrigerator has logged and takes pictures of everything that's inside there. It knows the expiration date of those things and a lot of times it knows when you put it in there, and then it says "OK with what you have today you can make a quiche that will feed six. And here are the seventeen things in your refrigerator you need to get out. Let me preheat the oven because it's going to take 45 minutes to cook this.

To Hickman, the attractions of voice-driven technologies are clear. "When the whole world revolves around the kitchen these days," he says, it's easy to find examples of speech-connected appliances' value. "So why does your stove need to talk to your Nest Protect home alarms? Well, that's because the Nest Protect [can] sense smoke but knows that there's food in the oven that that's been in there for an hour and a half. Not only can [it] sense that it's food smoke and not flames smoke or plastic smoke, but it could tell you that you need to pick up food on the way home 'cause someone burned dinner again."

NMG's special advantage, Hickman contends, is training at the dealer-member level. The local Home Depot, he argues, is not likely to have a salesperson trained to help a customer see the value of a truly connected home. But at "your local independent dealer . . . they have all those products displayed. They have an educated, directed salesforce" trained by NMG to carry out the strategy. If someone comes to a store "that has these products on their display and goes 'Good Lord! What's all this connected home thing,' That's where the value proposition starts to play out." NMG is training salespeople to cultivate attachment to voice-driven devices. "[The salesperson says,] 'Can I show you how to make your life easier? You know my wife can start our dryer from her car so that when she gets home, she folds clothes that are recently dried, not wrinkled, as opposed to coming home, realizing that they're wrinkled. Starting to make dinner, she has to go back and start the dryer again and then come back, interrupting dinner, to fold those clothes.' There's plenty in there," Hickman concludes. "There's some [benefits] I'm sure we haven't even thought of. [Those are] examples of real life application that we use today."

Most appliance firms are making their products compatible with Alexa, Google Assistant, or both. Samsung, a major appliance maker, allows its smart appliances to work with both, often through its SmartThings hub, despite offering its own Bixby voice assistant. To encourage customers to use Bixby, Samsung during 2019 announced Bixby2, a reboot of its voice assistant aimed at gaining parity with Google on Samsung phones as well as on its refrigerators and other appliances. The company showed its eagerness to push its avatar into the top tier in the United States when it hired one of the creators of Siri, Adam Cheyer, to lead the work.[17] Still, in mid-2019 Tom Hickman didn't seem aware of Samsung's grand plans for Bixby, and when I told him, he doubted it would get great traction. The public doesn't recognize Bixby, he asserted, and Samsung is just "one manufacturer versus Google and Alexa," which are "almost ubiquitous with whom they operate." NMG endorses Alexa and Google Assistant.

Toward the end of our interview, I noted that there are marketers who want to learn "what people say through those devices and then market back to them based on what we find out." Hickman responded that he "could absolutely see that being a way to delve into the hearts and minds in the intent of a consumer's journey, but I think it's got a lot of challenges in application." When I noted that Google's current privacy policy doesn't preclude this sort of activity, he said experiencing it would be "a wakeup call for that person who wasn't really paying attention to the privacy policy." He went on to suggest that this development is an opt-in, opt-out question for the people who purchase appliances rather than a problem that NMG had to address with its retailers. It's clear that Hickman sees NMG's retailers as presenting customers with attractive, connected-home options, and believes customers have to decide for themselves if there are risks.

The notion that the customer must take on this responsibility can also be seen on the website of one of NMG's most upscale corporate partners, Control4, which goes beyond kitchen and laundry machines to create high-end voice-enabled technology for alarm monitoring, audio/video media, and other connected home capabilities. The company represents Brad Russell's third level of customer adoption: major home renovations or custom new-home designs that emphasize voice tech. A video on the Control4 website (owned by Wirepath Home Systems) seductively depicts the voice-led connected home. It tells potential customers that "smart homes are the most meaningful when every technology inside works together to simplify your life." The video shows household members clicking on a phone app and an iPad-shaped house-control panel to activate lights and music, with the narrator touting the ability to get "more comfort and convenience using only the sound of your voice" as the camera pans to an Amazon Echo. Control4 emphasizes that it can connect a huge range of products to its system, displaying logos that include Apple TV, Roku, Lutron, LG, Honeywell, Yale locks, Kwikset, Sonos, and Samsung.[18] But if you want to know what happens to your data, Control4 sends you to vague, assuring statements accessed through a small-print link tagged "privacy and security." If you dig deeper by finding and reading the company's dense End User License Agreement, you'll learn that Control4 "disclaims all liability of any kind of Control4's suppliers [that is, any of the above connected-home firms] . . . [and is] not responsible for any liability arising out of content provided by . . . a third party that is accessed through Control4."[19] Control4 can hardly be unaware that Amazon and other companies may exploit the data that people generate when they speak, but it assumes no responsibility—even though, as a contractor, it is letting those firms into the home.

The same attitude shows up in the most seamless insertion of voice products into homes: the work of the home-construction industry. In their competition over customers, builders simultaneously seduce people into using the devices and ease habituation to them. Longtime construction leaders such as Lennar, KB Homes, Shea Homes, Brookfield Residential, Pulte Group, and Toll Brothers put the installation of connected-home technologies into overdrive beginning around 2018, when they shifted from experimentation with suites of the items in a few markets to standardized rollouts in most or all of their locations. The companies clearly saw connected homes as a way to attract buyers to their brands, and they promoted their association with Amazon and Google as assurance that the devices would work. Toll Brothers announced in a press release that as of July 1, 2018, "all homes entering into a sales agreement will include keyless entry by Baldwin, Wi-Fi thermostats by Carrier, and Wi-Fi garage control, along with either a Wi-Fi prewire or full Wi-Fi package (select areas)." It promised Amazon and Google voice compatibility, and gave customers the option to buy additional packages from Control4 and Alarm. com.[20] A couple of months later, KB announced that it had partnered with Google to bring connected home technology to newly built properties in Orange County, California; Las Vegas, Nevada; Denver, Colorado; and Jacksonville, Florida. "The KB Smart Home System includes: two Google WiFi points to create a mesh network, two smart speakers with the Google Assistant built-in (a Google Home and Google Home Mini), a Nest Hello video doorbell, and installation help." The result, according to the KB website, would be "Home Life, Uninterrupted," and it would be hands free: "Just start with 'Hey Google.'"[21] Not to be outdone, Lennar teamed up with Amazon's Home Services division to build its standard smart home around Alexa. Having confirmed with an organization called the WiFi Alliance that its construction setup would yield excellent signal reception throughout its houses, Lennar billed its offering as "the world's first WiFi certified home design with activation and support by Amazon."[22] Each home in summer 2020 came with "smart home products from the most trusted manufacturers, including Ring, Honeywell, Lutron, Ruckus, Schlage, and Sonos, [as well as] Amazon expert home activation and support."[23] Lennar emphasized that the house's network technology would enable "buyers to control their lights, front door lock and thermostat with hands-free voice commands to Alexa."[24]

The builders' choices of technologies to offer were arrived at only after a great deal of deliberation, and they were based on decisions about what voice-first devices would both attract their target customers and be easy for them to use. Tony Callahan, vice president of corporate purchasing for Shea Homes, told me that the firm's decisions about the makeup of its connected home were made with the help of many interviews with prospective house buyers.[25] The firm first hoped for compatibility with Amazon, Google, Apple, and Samsung products, but concluded that while compatibility with one firm was not difficult, harmonizing with all four "was tough." That was especially true of Apple, because compatibility sometimes required that an Apple chip be embedded in the product. As it turned out, the high visibility of Amazon and Google's voice devices guided Shea's solution. "As we looked at our market research," Callahan said, "we heard Amazon the most from our customers, then Google. We didn't hear much about Apple and [not] at all about Samsung."[26] Using guidance from this research, Shea's executives defined a connected home as having a WiFi router, whole home WiFi coverage, a connected deadbolt for the front door, a WiFi garage door opener, WiFi light switches, a video front doorbell, a WiFi thermostat, an Echo Show 5 to control the devices, and (Callahan said, reflecting a phrase from Lennar) "white glove service so it's connected for the home buyer." In some markets, the Shea's base package also includes connected appliances rather than standard ones.

The cost of implementing these technologies, Callahan noted, is very high. "I can't quote you that number, but it's a big deal. I mean, that's the reason more builders haven't done it. . . . We took a big hit. I can tell you there is no [profit] margin. We're doing it at cost, and we're doing it because we think it will help us sell homes. But this is not something we are doing because we think we're going to make money. . . . We talk to prospective homeowners. They want this." Asked if Shea's research found security and entertainment to be people's dominant motivations, Callahan said, "No, it's about convenience."

> You get home from a long day at work—both spouses are working—and you're helping the kid with the homework and you want to set the oven timer [to delay cooking until a certain time], so you don't have to get up and go check something in the oven. You can do that from your phone or your voice controller.
>
> You are running a little late—the kid still doing that homework—and so you can delay your oven timer. You know you're going to be taking a shower so you start the recirc pump [via voice command] so as soon as you turn the shower on you have hot water. You're going to bed at night. You say "Alexa, goodnight" and it locks your front door and turns off all your light switches. [You then worry,] "Man, did I lower the garage door?" Well, you look at your app and you say "Yup, I lowered my garage door," or "No I didn't: lower it!"
>
> It is convenience.

When asked if they had considered using any of the data flowing through customers' homes for their own purposes, home builders responded indignantly that they are in the construction business, not the customer data business. Yet the home retailers and builders I interviewed didn't bring up surveillance concerns as factors in decisions about the appliances they were installing in people's homes. Apple's attempt to convince the world that it is more privacy-friendly than its rivals doesn't appear to have been a factor in the builders' competitive assessment about what technology to package for customers.

If the home is a prized location for getting people into the practice of using voice assistants, the car is a close second. The amount of data potentially available from people's interactions with their cars is huge and often quite personal. The website Mashable wrote in 2018 that "Voice has become the go-to tool for the modern household, such as in the 'smart' kitchen or living room, and even more so in the car." In fact, it added, a 2017 survey from the digital consulting firm Capgemini found that 85 percent of respondents "prefer to use voice assistants on the go"—that is, in the car, while biking, or on the train. That number was similar to the 88 percent who preferred to use voice assistants in the living room.[27] It's not hard to understand, then, that the car has become a central stage of competition among voice-first firms to convince people that sharing their needs with a talking computer is often not just convenient, but also the safe thing to do. Car manufacturers are aiding this habituation process.

In 2019 the voice-recognition firm Nuance underscored the growing importance of voice competition around cars by spinning off its automotive division into a separate company, Cerence. Nuance has long been the dominant force in making speaking to cars a habit, with Google, Apple, SoundHound, and Amazon lagging behind. "We've been working with the carmakers for over twenty years," said Eric Montague, the company's automotive strategist, in 2019. "Today, every major carmaker is using Nuance technology in one form or the other. So we work with over fifty different brand of cars. I'm sure you couldn't even name fifty brands."[28] Asked the purpose of a car's voice technology, he said it was "to enable drivers to operate a vehicle more safely." Voice activation is easier to use in a moving car than push-button interfaces. Nuance in 2019 was marketing sixteen different mobility assistant modules to carmakers, or what Montague calls "original equipment manufacturers" (OEMs). A mid-range brand would buy two or three. A high-end brand might buy all sixteen. "A lot of OEMs," he said, "want to bring new innovations first to infotainment systems in the vehicle to differentiate against their competitors, or because certain brands are defined by the level of product innovation that they bring. Mercedes, for example— . . . they always want to be first in new innovations, new functions, new features." As if to elaborate on the point, a Mercedes Benz senior software engineer noted at a 2019 conference for voice-app developers that his company had decided to place some of Nuance's most cutting-edge voice control innovations into its entry-level 200-series models. The reason was strategic. Company research had found that younger adults, the people most likely to buy an entry-level model, were substantially more likely to want the latest tech and know what to do with it. Because Mercedes executives felt the company needed to attract more young drivers, Mercedes filled its 200-series cars with voice tech beyond what customers might expect.[29]

Cerence acts as a white label provider with its auto clients. Largely unknown to drivers, its role extends deeply into the operating systems of vehicles, allowing it to retrieve and process enormous amounts of data for and about the people in the car. At the 2019 Consumer Electronics Show, Cerence presented a software package it is refining that implements the driver's complex interactions with the car via speech, voice, facial expressions, and vision. Called Cerence Drive (before the merger it was Dragon Drive), the advanced mobility assistant is based on the proposition that "operating the car without a need to push a button, use a rotary switch, or even touch a screen is not only possible but significantly improves the user experience." It makes all of this possible by taking the multi-array microphone setup, gaze detection, and gesturing that are already available in expensive models to an even more sophisticated level. Using voice commands and eye movements, drivers can interact with symbols ("widgets") displayed on the windshield "to refine and select services and information—phone, contacts, weather, navigation, music"—that have traditionally appeared on the car's center console. A Nuance press release from before the merger points out that the Drive's ability to sense both head movement and voice allows for touch-free in-car controls. For example, the driver can look at the passenger-side window, say "Open that window," and it will happen. The press release further notes that beyond understanding what drivers and passengers say, "Dragon Drive now also senses how they feel"—their "cognitive and emotional states." The result of a joint effort between Nuance and the Massachusetts-based startup Affectiva, this feature uses cameras to register the driver's facial expressions and the car's microphones to discern tone of voice. Affectiva's "Emotion AI" software informs the mobility assistant of drivers' and passengers' cognitive and emotional states. From there, "the assistant adapts its behavior accordingly, changing both its response style and tone of voice to match the situation. Besides providing more 'empathic' assistance, this technology could enhance safety on the road by preventing distracted, drowsy and impaired driving."[30]

Although these capabilities seem formidable, Cerence's impressive hold on the car's voice is eroding in favor of Siri, Google Assistant, and Amazon. Although it offers a great deal of expertise with car instructions, the company doesn't have the wherewithal to connect consumers with the world outside the automobile, as Google, Apple, and Amazon can do through their voice assistants. And Cerence's impressive features don't satisfy the increasing proportion of drivers who want to do through their car's display and sound system what they do with their smartphones and home speakers. A 2019 study by JD Power that was supported by Amazon showed that "59% of U.S. consumers said they were more likely to purchase a car from auto brands that support their preferred voice assistant used in the home."[31] Amazon's "chief evangelist" for Alexa automotive took this to mean that home-targeted habituation had successfully taught people to want the same voice assistant outside the home. "We've reached that tipping point in voice," she said.[32]

Many car manufacturers have jumped on the bandwagon, hiring Cerence or another firm to enable drivers to port the voice and touch features they routinely use from Apple, Google, and Amazon to the car's microphones, speakers, and display. The calling capabilities, dynamic maps, music apps, and information-and-entertainment features that people see on their car screens at this point merely mirror those of the phones they carry with them, rather than being integrated into the depths of the car's voice-activated system. "Apple Carplay and [Google's] Android Auto are available in even the cheapest models now," said Ed Kim, an executive at the automotive research firm AutoPacific. Not to be left out, Amazon, despite its presence on many fewer phones than

Google and Apple, entered the passenger cabin with its Echo Auto module in 2017, a few years later than Apple's and Google's car-mirroring tech launched.[33] In mid-2019, the company was selling Amazon Auto for the remarkably low price of $25, presumably to compete with Google's and Apple's car offerings, which cost nothing to download.

Many carmakers were beginning to see a long-term advantage to bringing Google, Apple, or Amazon more deeply into their systems. Nuance's chief technology officer, Joe Petro, explained that the OEMs are looking ahead to self-driving cars and are beginning to think of the cabin as an information-and-entertainment space with the sounds and sights that customers find comfortable. Their interest in established voice tech is an indicator of the voice-first titans' effectiveness in habituating large numbers of Americans to their devices. A related indicator is General Motors' announcement in late 2019 that it would offer owners of 2018 and later models the ability to download Alexa as an integrated add-on to the company's own voice system.[34] A broader solution to the perceived need to allow drivers' habitual voice assistants into the car is the multi-assistant capability rolled out in late 2019 by Amazon and Nuance, which allows the driver to seamlessly call up whatever assistant he or she prefers. You can speak to Alexa or to the BMW assistant, for example, without skipping a beat. Google and Apple didn't immediately join Amazon in this project, but voice-industry expert Bret Kinsella predicted that "consumer expectations" will eventually "make it harder for Google or Apple to enforce single assistant policies."[35]

Yet some car companies—such as Mercedes—want their customers to see their brand as the dominant voice assistant for everything in the car, even as they recognize that their interfaces don't have the generalized knowledge of a Google Assistant, Siri, or Alexa. Joe Petro pointed to Nuance's (now Cerence's) ability to quietly integrate this multi-assistant solution through "cognitive arbitration," a process for deciding which agents the Nuance software should call on for desired information or entertainment from among the providers the OEM permits.[36] So, said Petro, "if you say, 'Hey Mercedes, what's the weather tomorrow,' it will go through the cognitive arbitration layer, and the cognitive arbitration layer will say, 'Aha! I'm going to bring this to Alexa,' or 'I'm going to bring this one to Google.'" Or Nuance may draw on Microsoft's Bing, which Amazon uses as its search engine, for the data. The passenger, Petro noted, will experience only the Mercedes interface.

Petro and Montague both emphasized that all of these activities raise issues of sharing data, including a person's speech, among the actors dealing with car information. The voice activation software and cognitive arbitration procedures are so complex that while the system answers some commands in the car, other commands must travel to the cloud to be recognized and addressed. "It depends on the command," Montague says. "If you control your car [say, tell it to lower a window or ask it for tire pressures], what you say to the car stays. If you want to search for something . . . it's the cloud that connects what you say to the content . . . you're looking for. So for example, if you say to the car, 'Find the cheapest [garage] near Times Square that has space,' then it would get that utterance to the cloud. It will be converted into text. It would search for providers to find which car parks have space available for two hours, and give you back the answer.

That happens in the cloud." The cloud would also allow the automaker to keep data about what individual drivers have wanted in their cars based on their voice commands. If a driver buys or rents another model of the same brand, the setup would be ready.

This level of sophistication in driver understanding is only the beginning. At the 2020 Consumer Electronics Show, Mercedes exhibited a system that recognizes the driver by breathing and heartbeat.[37] Amazon's booth showed off "home-to-car integrations with Alexa and Ring, in-car Alexa voice features, and [Amazon cloud] services that can help automakers and car dealers bring more technology" to the car.[38] "A car is now listening to everybody" inside it, asserted Petro, mostly so that Nuance can distinguish the driver's voice from those of the others in the vehicle. "Because when the driver is controlling the car, the driver is the dominant personality, and from a safety point of view the driver is the only person who can lower their window, set the GPS, so kids screaming in the back need to be screened out of that." Moreover, Petro says, the latest Nuance car-mobility assistant can, with the help of a manufacturer-installed camera and software development kit, detect where a driver is looking and respond to requests based on that. He gives, as an example, the driver saying "What time is that open?" while looking at a store on the left. Hearing the question, Nuance "queries the car video system" about the direction in which the driver is looking. Nuance's cognitive arbitration process then sends the question to the manufacturer's provider of GPS and mapping, which comes up with the answer and hands it off to the Nuance speech system, which tells the driver. "We call that conversational AI," Petro noted. "The gaze detection, gesturing, and multi-array mic [are] all part of the car now."

Clearly, what people say in a car, when and where they say it, and with what emotions are all of great potential value in creating profiles for marketing and other purposes. Both Eric Montague and Joe Petro emphasized that Nuance handles voice information only for the specific in-car experience to which it relates. Apart from that, Petro said, "we give control over how data is used completely to the carmaker." He said that Nuance is completely open with the carmakers about its deployment of the voice information. "If we're going to use data to train something, our clients know exactly what we're using the data for, and in a general sense it's only used to quote unquote improve the system." Asked if the auto manufacturers keep a record of what people do and say in their cars, he answered, "Only if you authorize them to do so. The default is not. You have to agree to terms of service." Customers may have an incentive to allow that authorization; buyers of newer cars are told that the more the car assistants know about them, the more personalized the responses will be. Asked if the car companies that Nuance deals with use the data beyond the particular instances of data creation, Petro answered: "for the most part everybody is trying to behave as well as they can."

But given the amount of driver data that is created every day, the temptation to use it is huge. Roger Lanctot, an automotive specialist at the consulting firm Strategic Analytics, notes that car manufacturers have only begun to think through what they want to do with the deluge of information. The most obvious approach, he says, would be to analyze the habits and interests of car owners in order to best understand how to get them to purchase future models. Customer retention is a critical goal for carmakers, and they will certainly not resist the temptation to use voice assistant technologies for these sorts of analyses. Using the data for other purposes is more controversial. Lanctot tells a story of an auto executive who suggested using driver data as a separate business line and whose superiors, sounding much like the home builders I talked with, told him, "We make cars. We are not in the data sales business." But as Lanctot points out, automakers do have a history of monetizing their car buyers. Some have rented buyers' names to tire, motor oil, and other companies as customer leads. Others have charged insurance companies for information about the driving habits of those owners who consent to the data flow in the hope of getting their rates reduced. Second-by-second information flows about where people drive, what they say, and how they say it offer an opportunity for car companies to create personalized audio or video ads, sell subscription services based on the data, or "tip off third party businesses" with leads based on very specific information.[39] As a 2018 *Automotive News Europe* article pointed out, "smart infotainment systems are revenue streams for automakers, creating a marketplace for third-party apps and services that, for instance, allow drivers to find parking or pay for a gasoline fill-up." The article quoted Sam Abuelsamid, senior analyst with Navigant Research, who said "Whoever owns the platform can skim a percentage off the top. Those nickels and dimes eventually add up to real money." Understanding what people are doing in the vehicle, he added, allows automakers to create revenue streams and improve the user experience.[40]

Major car manufacturers haven't discounted that possibility. BMW admitted that it tracks its drivers but made it sound altruistic. "Let's say a person is listening to certain music, and we know there's a big concert," said the senior vice president of digital products, "Then we would probably give that to our salespeople to make an offer for a special ticket."[41] A Volvo statement obliquely conceded using driver data, telling a Fortune.com writer that the company's technology "takes full account of legal, security, and privacy obligations on a global scale" and complies with a European Union law that lets residents control how their personal data is shared.[42] A Google online search for *speech* or *voice* and Volvo *privacy policy* during July 2020 did yield an "Information Notice— Speech Messaging" that (consistent with EU regulations) noted that when a driver talks to the car "Volvo uses a 3rd party service provider to perform the voice recognition and also to improve the service provider's service."[43] The notice did not reflect the broader comment made to Fortune. A similar search for a BMW or Ford privacy policy on the topic turned up nothing. Similarly not useful was a phone call to BMW's customer service and "genius" advisers. A genius adviser said that buyers sign a privacy disclosure sheet at the time of purchase, but he could not point to a place on the web where a person could read it. Agreeing that it's difficult to find information, Lanctot told me that the companies' approach to their sophisticated voice assistants is still in its early days. He suggested that if executives hope to make money from the data through activities such as ad personalization, a primary concern they are confronting is "how to do it without being too intrusive" in the eyes of restive regulators and concerned publics.[44]

Whether or not OEMs use drivers' voice data for marketing, they certainly allow Google, Amazon, and Apple to do it. Auto industry analysts understand that for the major voice firms, connecting with individuals in their cars means being able to continue tracking and profiling them throughout more of their daily lives. According to the AAA Foundation for Traffic Safety, the average American driver spent fifty-one minutes a day in the car during 2016–2017, the most recent year it gathered the data.[45] For Apple, the dominant reason for a helpful connection to the car might simply be to encourage iPhone purchases. Apple's browser targets individuals for ads, as do many of the apps its phones carry. Similarly, Amazon's interest, Lanctot suggested, is in getting people to purchase things from their car by voice— either while moving or while seated in a stopped or parked vehicle. "The car is a browser's last frontier," he said.[46] Moreover, the driver and passengers are "seat-belted in place"—"a literal captive audience," in the words of the *Wall Street Journal*.[47] With Google, Apple, and Amazon treating the automobile as a potential agent of its operating system, the *Journal* painted a future scenario in which voice, when combined with features of the car's video screen (especially its real-time maps), would enable heavy cross-device targeting and selling:

> On future screens, local restaurants, doctors' offices and other services could target ads based on typical driving routes. An insurance company could offer lower rates for cautious drivers, while car makers could use system data to offer service on an aging part before it blows. Some envision a world where users could start watching a TV show at home, then with a voice command continue watching the same program in the car. Others are working on allowing users to order and pay for gasoline and coffee on their screens.[48]

Two people with deep knowledge of the auto industry volunteered that for Google, even this level of knowledge doesn't seem to be enough. Recently the company has been working on a personalized voice system that offers deep integration similar to the (typically Cerence-created) technologies of the car manufacturers. The move has caused a bit of a firestorm. An executive who wanted anonymity noted two problems that are dissuading some carmakers from adopting it.[49] One is the issue of branding. According to the executive, Google, unlike Cerence, insists that the product be labeled Google and that drivers say "Hey, Google" (instead of a car brand) when they activate it. Executives worry that while this helps Google, it may devalue the carmaker as the curator and manager of a vehicle's features for its occupants. But the second problem with adopting Google's integrated voice assistant is the amount of data the company would be able to collect. Automakers have essentially accepted that Google, Amazon, Apple, and their apps are learning about drivers' locations and spoken interests. Some automakers balk, however, at giving Google a portal into far more information about people's automotive lives—information that the manufacturers consider proprietary or invasive. One former BMW executive told the *Wall Street Journal* that Google executives asked if BMW could put sensors in the passenger seat to determine based on the occupant's weight whether a child or an adult was riding in the car. It's easy to imagine the executive assuming that this and more, including what people say and how they say it, could be used to target drivers and passengers in other places.[50]

Despite Google's obvious history of making almost all its money through personalized advertising, a number of carmakers, notably the Renault-Nissan-Mitsubishi alliance—one of the world's largest car-making groups by sales—entered into a partnership with the company "to eventually install its operating system in the 10.6 million cars the alliance sells globally each year." Volvo has also partnered with Google. The *Wall Street Journal* reported that Volvo employees suggested it wouldn't be good to hand over such an important system "to an outsider," but company leaders concluded that because Google's Android system would win in the marketplace, they had better sign on. Renault also had suspicions before it accepted a deal. Its CEO said, "We spent more than four years, almost five, to find an agreement with our new Google friend precisely to have control of the data." He declined to disclose what "control of the data" means.[51]

Eventually both Volvo and the Alliance's car privacy policies will have to reveal exactly what data Google will be using and whether voice will be part of it. Joe Petro, whose company obviously has a competitive interest in limiting Google's movement into car operating systems, noted that people receive those terms of use when they purchase the car and likely don't read them. He termed the lack of certainty about what information firms are using "spooky." Reflecting this distrust and nervousness, an auto executive I interviewed who wanted anonymity invoked a discussion he had with "somebody at Google" who he claimed told him that "if carmakers really knew what's behind Google, it would really scare them. They don't even know what Google is doing."[52]

So far, the home and the automobile are where manufacturers, retailers, installation firms, and app providers have most successfully attracted people to voice-driven technologies. But it's just a matter of time before voice-first technologies become more obvious in other sectors of American life as well. The manufacturers of smart speakers are pushing new angles for the technology. Independent innovators are working to transfer the benefits of voice-first devices in their homes and cars to their jobs. And marketing agencies are looking for ways to put voice gadgets in new places. In all of these situations, the positive, seductive qualities of voice tech are stressed while the all-important surveillance aspects are downplayed. The goal is to make voice-first thinking feel as natural in new venues like hotels, schools, and stores as it is becoming in homes and cars.

Take the hotel business, where Amazon and Google are working to spread smart speakers' influence. Amazon was first out of the gate. It launched its service in 2017 in collaboration with hotel giants like Best Western, Wynn Resorts, and Marriott. Google waited until 2020.[53] Amazon boasts that "Alexa for Hospitality," an initiative of Amazon Business, will make the smart speaker "the hub of the room, creating a guest-centric experience . . . that all guests can enjoy."[54] A video pitch for hotel executives shows a businessperson in her hotel room asking the Alexa Echo on the desk to turn on the lights, "play my focus playlist" (she clearly has an Echo at home), and order an Ahi tuna salad. When she requests the food, Alexa asks if she would like to add a glass of red wine, and the guest says, "Yes, sounds perfect." The video shows the woman sitting on the hotel room couch asking Alexa to wake her at 6:15 a.m. and to turn off the lights. In the morning, she says, "Alexa, clean my room," and Alexa responds, "I've notified house-keeping. Enjoy your day." Later, as she is packing, the woman says "Alexa, call home," and she speaks to her son, telling him she looks forward to seeing him. She finally says, "Alexa, help me check out," to which Alexa responds, "OK."[55]

One website commented that for hotel guests with home-based Alexa accounts, this "level of personalization lets weary and homesick travelers bring a piece of home with them to make their stays more comfortable."[56] Yet the hotel industry is divided on the installation. At a 2019 meeting of the Americas Lodging Investment Summit, Best Western Hotels & Resorts CEO David Kong told his audience that a pilot program using Amazon's cylinder in Best Western rooms "did not go well. . . . When most people got into their hotel room, they disconnected it, presumably because they didn't want Alexa listening to them in the room." Usage was low, and "we didn't see any lift in satisfaction scores."[57] Kong added that he also received complaints about Alexa activating unprompted in the middle of the night and waking guests. When asked by a panel moderator if he would unplug a voice-activated virtual assistant device in his own hotel room, Kong said he would.

But according to the *Travel Weekly* website, some Echo-outfitted hotels have had better luck. One is the Westin Buffalo, which rolled out the device in 2017 and now has one in each of its 116 rooms. The hotel's general manager of property said that the Echo offers "guests an easy, fun way to request services, get recommendations and generally engage with hotel staff without the hassle of picking up a phone." Moreover, "from the hotel perspective, seeing the most common requests has allowed us to better anticipate our guests' needs and therefore provide better service. Concerns about privacy have been minimal, [but] in the event of a concern, we recommend the guest turn the device off or unplug it." Another professional observer of the hotel industry saw guest concerns as a problem to be solved. "The jury's still out on the true application and success of voice in hospitality. I do think there's a place for it, but we ultimately still need to figure out the logistics around it."[58]

Amazon hasn't been deterred. It's doubling down by trying to persuade hotels that putting Alexa in their rooms not only will enhance guests' experience, but also can improve the hotel's efficiency. That is, the pitch to hotel executives pushes both personalization for the guests as well as the ability to organize customer data in unprecedented ways. "Your guests want hospitality experiences that are personalized, and memorable," the video says. It stresses that the hotel will save money by using Alexa as a "virtual concierge," and that "you now have immediate access to guest engagement and feedback, helping you adapt your hotel experience to meet your guests' needs."[59]

Google's approach is similar to Amazon's. Its product manager writes that "From a Nest Hub set up in each hotel room, hotels can tailor the guest experience with Google to specifically address common service requests from guests through a simple voice command."[60] Google and Amazon are not alone. Angie Hospitality also offers a virtual concierge. It leases or sells three devices specifically built for guest interactions with an increasingly sophisticated range of capabilities, from concierge interactions to music, telephone-related tasks, heating, checkout, and more.[61] What Amazon and Google are selling, though, is expertise in AI-driven speech recognition along with the familiarity many guests already have with Alexa and Nest, despite some who are suspicious of having voice-activated devices in their hotel rooms.

Volara is a company in the business of convincing hotels to let it install smart speakers. At first it used only the Amazon Echo, but in 2020 Volara began to use the Google Nest Hub as well.[62] Its CEO, Dave Berger, severely critiqued David Kong's device rollout, arguing that strong WiFi and customized devices needed to be installed first.[63] He contends that hotels will find increasing numbers of guests attracted to the devices as more and more Americans have smart speakers at home. Using Amazon's software developer kit, for example, Volara has created a system where a guest can ask Alexa general questions that one might ask at home, whether or not the guest has an Echo account. But the guest can also ask questions about the hotel and its services that Volara in most cases anticipates by using machine learning to continually adapt to what hotel visitors want to know. Berger says the questions follow the 80-20 rule; its system can answer 80 percent of the questions directly. If a person asks for extra towels or food—two reasons guests typically call room service—Volara sends requests to the work-order management system, which is integrated with the hotel's computers. In the other 20 percent of the cases, it does not understand what the guest wants so the request is sent from the smart speaker to the relevant person in the hotel.[64]

Asked about the confidentiality of the data, Berger said that neither Volara nor the hotel receives a voiceprint or full transcript of questions that the guest is asking of the hotel—just phrases or words that indicate the intent. He compared it to hotels keeping records of the requests that guests make when they call the registration desk or concierge. Amazon itself emphasizes, in a "Question and Answer" sheet created for reporters writing about the service, that hotel properties "can't listen to what guests say to Alexa or what she says back." But the Q&A says nothing about *Amazon's* use of what guests say to Alexa. Its privacy policy doesn't stop Amazon from collecting what guests say when they sign in with their Echo accounts to use their personal music settings and phone contacts in the hotel. The policy doesn't even prevent Amazon from gathering what guests say, giving them cookie-like tags, and tracking them across hotel stays and elsewhere.[65]

While Amazon is actively trying to spread the Echo into hotels, its diffusion into classroom learning so far seems driven primarily by teachers whose home experience with the device has convinced them of its classroom utility. Consider Ask My Class, an entrepreneurial elementary-school project backed by the Y Combinator venture capitalist fund. Aparna Ramanathan, an anesthesiologist, started the operation after seeing her children interact with Alexa at home. She liked that it can be screen free and believed she could make money by creating apps that empha-size voice games for children.[66] After trying and failing to get fami-lies interested, she decided to focus on elementary schools. Some teachers, she saw, were already using the Echo in the classroom as activity timers and for other simple purposes. Confident she could add value, she and her tech-oriented husband created Echo education modules beginning in late 2017. By late 2019, around a thousand teachers from throughout the United States were paying twenty dollars a month to access a wide range of voice materials for classrooms. Ramanathan tells me she eventually hopes to license the modules to entire schools rather than indi-vidual teachers.[67]

The AskMyClass website describes its product as "the ultimate extra pair of hands for teachers," with "100s of ready-to-use activities and games, at your command." The company has two Alexa-based features, one that provides activity modules for the classroom and the other, called "class picker," that can randomly call out students by their first names so that students don't think the teacher is playing favorites. The activity modules are one to three minutes long and range from "countdown and timer" announcements ("You've got one minute, starting now [bell rings]"), to "transition time" ("Pay close attention: If you walked to school this morning, get ready for dismissal"), to "emotional support" ("If you're feeling angry, frustrated, or mad, then you're in the red. Let me try and help. Stay in this quiet place with me").[68]

Others are also trying to migrate smart devices to the school environment. According to *Education Week* magazine, the spread of voice-first technology into American homes has made "many children . . . as comfortable giving verbal instructions to an Amazon Echo or Google Home device as they are talking to a person."[69] This development is "forcing educators to think more deeply about the role they play in developing the technological fluency necessary for students to communicate effectively with artificially intelligent machines." Some teachers celebrate the immediacy of the smart speaker and argue that learning how to use it outside the home will teach students skills they will need in future work environments. Based on research in four K–12 school districts in Idaho, an associate professor of education at the University of Idaho "witnessed a wide range of uses, from helping students with learning disabilities develop communication skills to getting students excited about specific research projects."[70] Detractors don't disagree that such benefits can accrue, but they worry about keeping what children say private. The U.S.

Department of Education lists situations where recordings from voice assistants in the classroom would violate federal student privacy laws—for example, an audio recording of a student being disciplined or having a medical emergency. Ramanathan says she is aware of the need to protect student privacy. That's why, she says, her Alexa-based technology talks to students more than inviting interactions; teachers might feel uneasy having children speak to a device that could capture their voices.[71] A few of her modules do involve kids speaking anonymously back to Alexa (for example, filling in a number-skip game), but in order for teachers to use them, parents have to give their consent.

Lost in the pro and con debate, however, is that teachers themselves are part of a voice collection trail. Amazon receives the voiceprints of all the requests and statements that teachers make (as it does with all adult users), and because every teacher must have an Echo account to use Alexa and AskMyClass, it can then analyze them according to that teacher's profile. In addition to enhancing Amazon's ability to target individual teachers in their private lives, the data can provide the company with a trove of knowledge about what its adult users say in classroom locations, knowledge that it could use to its advantage.

The diffusion of smart speakers has started in earnest. In early 2020, 11 percent of educators responding to an *Education Week* survey said their school districts are using voice technologies such as Amazon Alexa and Google Home "a lot" or "some" for teaching and learning.[72] Responding to the enthusiasm of voice-first tech in U.S. schools, Emerson College now offers a professional studies certificate program in voice design, and it will soon allow undergraduates to minor in the topic. The college has received an Alexa

Fellowship from Amazon, which provides funding and other help "to invent the next big thing in conversational AI."[73]

By *Education Week*'s account, surveillance is generally an unspoken, accepted part of the atmosphere fostering smart-device diffusion in many classrooms. Critics caution that the lure of voice-first education is not being matched by admonitions about how devices can monitor individuals' activities, especially to the students who are learning from parents and teachers that giving up their voice to technologies is a useful part of life. "I don't see any evidence that students are being 'taught' here how to use these devices appropriately or warned of their risks," Leonie Haimson, a student-data privacy advocate, told *Education Week*. "Instead, it appears that there is a push to condition students early to accept the inevitability of surveillance and the violation of privacy it entails, as well as the mechanization of their classroom experiences, rather than resist this."[74]

If playing up seduction and playing down surveillance are helping the voice intelligence industry to penetrate schools and hotels, it would be odd not to see the same tendency in retail establishments. Over the past decade, stores have placed in their aisles a variety of Bluetooth, sound, video, and light technologies aimed at tracking customers as they shop. Adding devices that speak with customers to infer their dispositions, personalities, or other characteristics is a logical next step in personalized selling. And ways to do that are coming together, though at this point they benefit Amazon, Google, and Apple more than store owners.

Many people are already in the habit of talking to Siri, Google Assistant, and Amazon on their smartphones while they shop.

And many use the Amazon app or their browsers to search for product ratings and compare prices as they make decisions in Bed Bath & Beyond, Best Buy, and other places. Amazon is vague about what it does with these conversations. A person who clicks on its device's microphone for the first time is notified that "Amazon processes and retains audio, interactions, and other data in the cloud to provide and improve our services." If you go to the privacy policy, you will learn that Amazon has the ability to identify your location as well as a unique identifier for the device that is with you. In the "Examples of Information Collected" section of its privacy policy, Amazon also states that "you provide information to us when you . . . talk to or otherwise interact with our Alexa Voice service." It adds broadly that this activity might result "in your supplying us information such as . . . voice record-ings when you speak to Alexa"—which implies that everything on the recording is considered information collected.[75] Elsewhere in the privacy policy, the company notes that it provides "ad companies with information that allows them to serve you with more useful and relevant Amazon ads and to measure their effec-tiveness." Although Amazon will not give the ad firms access to your name or address, it claims the right to help those companies use data about where you are, as well as information that Amazon has inferred about you over time (logically, including information from your voice), to target you with personalized ads.[76] This type of personalization is also carried out by Google. For both compa-nies, what individual shoppers learn about products or what discounts get highlighted may now or in the future be based partly on what those shoppers said to their devices and how they said it.

As mentioned earlier, this setup currently benefits Amazon and Google more than the brick-and-mortar stores where people shop. A few "shopper marketing" agencies—companies that try to figure out ways to sell to people as they shop—are trying to change that by persuading stores to use smart speakers to interact directly with customers from shelves or displays. As Patrick Givens of Vayner Media describes the activity, it typically begins with a decision to use the Alexa for Business functionality on an Echo.[77] That allows a company to lock out typical uses of Alexa and focus users' attention on a voice app that pertains to the company's interest—for instance, in a store, interacting with customers around the benefits of the product. Probably the first instance of this activity took place in February 2018 at the Bottlerocket Wine & Spirit store in the Flatiron District of Manhattan. The voice assistant, designed to mimic a whisky expert, was implemented as a joint project of the Mars Agency, a Detroit-based shopper-marketing outfit, and SmartAisle, a subsidiary of Mars that aims to provide "a scalable, digital in-store assistant that augments human store associates, and provides shoppers with expert product selection, guidance, and information at the shelf." "Here's how it works," says SmartAisle's website. Using an Echo, the company's "intuitive voice interface engages shoppers in a natural conversation directly at the shelf." Next, "our custom, data-driven algorithm synthesizes shoppers' input/answers and determines the best set of recommendations" for purchase. Then "our responsive lights illuminate the featured products on the shelf, helping shoppers easily make their final selection(s)."[78]

At Bottlerocket, the Mars Agency's goal was to get people into the store to learn about whisky from the Echo so they would purchase what the app determined to be the best brand for them. The premise, stated in a press release by Bottlerocket's owner, is that "Buying whiskey can be challenging." A sidewalk sign outside the store read "If these shelves could talk . . . Actually, they CAN! SEE FOR YOURSELF." Inside, in front of a long right-angled shelf displaying a hundred bottles of whisky, a circular floor mat read "STAND HERE AND TALK TO THE SHELF (YES, REALLY)." An Echo tied to the display had a printed sheet above it urging the visitor to say "Alexa, open bottle genius." In a Mars Agency video, a visitor does just that. Alexa answers, asks whether the shopper is looking for "a gift, something similar to your favorite, or something different." The woman answers "a gift," Alexa replies that "whiskey makes an awesome gift," and the two then engage in a series of question-and-answer exchanges (which includes Alexa asking for a target price) that yields four suggestions and descriptions of each, with a light on the shelf highlighting each choice. The shopper chooses one of them, and Alexa says, "Thanks for hanging out. Let's hope the person getting the gift lets you have a sip or two. Come back and visit the Bottle Genius soon."[79]

Ethan Goodman, senior vice president of technology and innovation at the Mars Agency, said in a press release, "We believe voice technologies like the Bottle Genius skill powered by Smart Aisle will trump mobile as the preferred mode or interface for shoppers within the next five years."[80] The draw for store owners is that voice tech can help ease the high cost of labor and address the need for knowledgeable advisers in the store. Although the activity may seem gimmicky, Goodman said in an interview that his firm and Bottlerocket were very happy with the results. He said that in addition to the whiskey store implementation, his company had created an interactive voice-activated virtual adviser (this time with a Google device) that Estée Lauder used with its cosmetics line. Adding to enthusiasm about this new frontier of voice tech, Peter Peng, CEO of the Jetson Agency, announced in 2019 that his company had made a deal with the beverage distributor Remy to "bring voice technology to the center aisle," and that Jetson had recently signed a letter of intent with Procter & Gamble to do the same. "The thing is," he told me, "voice commerce can happen anywhere." People are currently "talking about it only in the home, but being able to do it on premises is the key."[81]

For now, in-store use of smart speakers is one case where marketers can legitimately play down the surveillance part, at least when it comes to individual profiling. One of the frustrations that Ethan Goodman had with the Bottlerocket trial was that his company knew nothing about the shopper standing in front of the device.[82] The person came in from the street, spoke to the device, bought a bottle of whiskey or didn't, and left. At the point of contact, each individual was anonymous, and the store wasn't able to go back to find out what the people whose names they did know (through credit card purchases) said to Alexa. Goodman said that his team at Mars has discussed future attempts to learn about the shoppers at the point of interaction so that they might vary Alexa's statements based on that intelligence. So far, he said, they're learning a lot about how much to prompt people about what to ask Alexa, and how to encourage interactions with the smart device.

The limiting factor that gives customers anonymity, at least for now, is that Amazon cannot identify particular Echo account holders fast enough. Voice business expert Brett Kinsella points out that the company can recognize voices in homes where it compares just a few known individuals. It can do it in hotels and cars when it knows the person whose voiceprint it must find because that person has logged into the system. But at this point, Amazon doesn't have the ability to sort quickly through billions of voice-prints to identify a particular individual.[83] That is likely to change, as Amazon scales up its voiceprint abilities. At present, though, the forced anonymity may help the diffusion of smart speakers across retail spaces, because neither the retailers nor Amazon needs to post notices that discourage shoppers from interacting with the devices. By the time Amazon and Google can identify a stranger talking to a smart speaker in a store, the habit of talking with these devices may be almost universal among shoppers.

The idea that the Echo or Google Assistant might become the operating system for your life turns out to be more real in light of Google's and Amazon's efforts—with help from building contractors, schools, hotels, and shopper agencies—to make this our collective future. Through seductive packages of handsome devices, very low prices (think Prime Day), and a gamut of institutional justifications (home conveniences, better schooling, a more knowledgeable concierge), they have taken on the mission of convincing Americans that voice-first instruments are wonderful conduits for personalized interactions with the world. It turns out, though, that surveillance in support of voice intelligence doesn't take place under only the industry's own seductive, comforting initiatives. As we will see, the media extend the reach of these surveillance activities in ways that grant the devices a legitimacy that the voice intelligence industry could never bestow on itself.

4 VOICE TECH
CONQUERS THE PRESS

"Tech is taking over your kingdom," CNBC anchor Michelle Caruso-Cabrera soberly informed her audience during a May 2018 spot on the program *Power Lunch*. "Amazon is teaming up with the nation's largest residential builder to sell and service smart home technology, and Diana Olick is live in North Virginia with more."

"Not creepy in the bathroom at all, though," says Olick, who is shown standing in a Lennar model house bathroom. Whether Olick meant "not creepy" to be ironic is unclear, but during the next two minutes she points enthusiastically to a parade of connected things, from lighting to window shades that residents can arrange through voice and touch directives, to a pantry that Amazon will replenish when the homeowner contacts the grocery store via the internet. Toward the end of the spot, Olick, sitting on a bed, notes that while Echo devices are visible throughout the home, Lennar has plans to integrate Amazon's voice assistant deep into ceilings and walls. She says she doesn't know if she is allowed to say exactly how Lennar will hide the devices, because it may be confidential. In the future, she says, Alexa will be "built into the home . . . like bugs."

A second *Power Lunch* anchor, Melissa Francis, interrupts: "I have two words, privacy concerns. There are no laws about that." Olick parries her. "I have one word, pantry. Pantry that gets filled. I mean, who cares if they know about what goes into the pantry as long as it comes to the door?" At which point Caruso-Cabrera ends the segment by concluding, "They're everywhere. Amazon is everywhere."[1]

What is the viewer supposed to make of this segment? The pat line "Amazon is everywhere" doesn't begin to resolve the tension between "privacy" and "pantry." It does imply that there will soon be no way to avoid Amazon's presence in, and even bugging of, the home. "Alexa, Should We Trust You?," the title of a November 2018 *Atlantic* article, better approaches the heart of the dilemma.[2] The plaintive and even subservient response— what could Alexa possibly answer but "yes"?—points to the article's theme: despite some serious drawbacks, the compelling features of Echo's voice assistant mean that it's probably here for good, whether we trust it or not.

We should hardly be shocked that Amazon and Google strategically create their promotions, prices, product designs, and relationships with home-improvement retailers, home-builders, car companies, and other industries in ways that make their voice-first devices easy for the public to want and get. What is disappointing is that the stories told by the media reinforce the desirability and legitimacy of voice-first devices and encourage our resignation that they are here for good whether we trust them or not. When Pinelopi Troullinou interviewed people in the United Kingdom about why they had adopted smartphones, she found that the media play an important role in promoting the desirability of technology. The people she spoke to had been "seduced by [media] discourses of convenience, efficiency and entertainment into handing over personal data" as a routine aspect of smartphone ownership.[3]

Beginning with Siri's launch, American media coverage has been much like what Troullinou found in Britain: largely devoted to cheerleading about convenience, efficiency, and entertainment. The occasional hand-wringing about surveillance through the devices is generally mixed with confusing assurances by the firms involved that it isn't as bad as some claim. Confronted with these contradictory messages, people may have stopped listening to negative press callouts about companies that are already an integral part of their daily lives. The voice-profiling powers benefit from this confusion. In an environment in which mainstream journalism is leading Americans toward resigned habituation with voice-first devices, trust is increasingly irrelevant.

When Apple released Siri in 2010, the popular press projected an enthusiasm about voice tech's amazing possibilities that set the tone for coverage ever since. *New York Times* reporter Jenna Wortham invoked the classic futuristic film *2001: A Space Odyssey* in an article about Siri's creator, SRI International, which sold the technology to Apple. SRI, Wortham stated, "is hoping to bring [that movie's] concept of virtual personal assistants closer to reality—without the malevolent malfunctions, of course." She marveled that Siri "allows people to perform Web searches by voice on a cellphone. Siri users can speak commands like 'find a table at an Italian restaurant for six at 8 tonight,' and the application can translate the request and use GPS functions and search algorithms to find an answer."[4] It turned out that Siri actually could not do that consistently, but the impression of great sophistication had been created. Another *New York Times* article assured readers that Siri's "personal assistant software is designed to listen to spoken questions, deduce the meaning, and act accordingly"—and that this was just the beginning.[5] One of Siri's creators, Norman Winarsky, told the *New York Times* that "Siri is the first and in some cases, the simplest [example], of what we'll do."[6]

Siri's arrival sparked a flurry of interest in voice assistants at the *Times*. Claire Cain Miller investigated what "Voice Actions," Google's early response to Siri, could do, and concluded that "Android cellphone screens could soon have fewer finger smudges."[7] A long *New York Times* piece by John Markoff and Steve Lohr reviewed research on interactive assistants that their creators hoped would one day replace physicians, car drivers, military translators, and contact center agents. Markoff and Lohr admired Siri's supposed restaurant capability "when it works," but allowed that it might not always work—one of just two comments in the article that cast any doubt on voice tech. The other was a critic's observation, seemingly aimed at contact centers, that "robot voices could be the perfect wall to protect institutions that don't want to deal with complaints."[8] In an indirect reply, Markoff and Lohr wrote approvingly of the ways that contact centers were focusing on complaints by analyzing callers' voices for emotions and treating them differently based on that data. A theme of the article was that voice-technology engineers were building a scientifically planned future without politics or social policies. They quoted Eric Horvitz, a Microsoft computer scientist, as saying, "Our young children and grandchildren will think it is completely natural to talk to machines that look at them and understand them."[9] Who wouldn't be attracted to such a system?

It wasn't just the *New York Times* that gushed about voice intelligence during the year following Siri's launch. Timothy Hay with the *Wall Street Journal* wrote that "recently . . . voice technology has turned a corner, and companies ranging from startups to tech giants are fighting to take advantage." This was all good, Hay explained. "While the field is crowded, the little guys aren't afraid of their larger rivals." Never mind that one of those rivals had actually bought Siri. Moreover, a little guy featured in the article, the Vlingo voice-search firm, would disappear two years later into Nuance. To the *Wall Street Journal*, though, market competition was working as it should. Importantly, Hay paraphrased a Nuance executive's assertion that the public was now responding to voice tech's sales pitch. "What has been lacking in the past—consumer demand—is now materializing."[10]

Other outlets picked up the claim that voice recognition had reached a tipping point. The *Philadelphia Inquirer* informed its readers that Siri was "an impressive advance in voice control and interactivity that some observers call revolutionary." It quoted a telecommunication industry analyst: "You basically don't have to type things anymore. You speak to your phone and it does it. You can ask your phone, 'Do I have time for lunch?' And then you can say, 'Reply back to Jeff that I have time.'"[11] Writers also reported that shoppers were responding enthusiastically to this technological breakthrough. A fall 2011 *USA Today* article noted that pre-orders and online orders for the new iPhone 4S with Siri were "backed up."[12] Another piece in *USA Today* said that "Siri's becoming everybody's buddy." Reporter Jefferson Graham marveled that Siri was "a sometimes sassy, often entertaining digital friend for millions of consumers" that "has, hidden in the software, dozens of humorous responses to questions. Consumers are asking Siri for dates, marriage, sexual advice, math equations and even crazy stuff like where to hide a dead body. To their surprise: Siri usually has an answer." The article pointed out that comedian Stephen Colbert had that week asked Siri, in jest, to write his show for him and that musician Jonathan Mann had written an online duet with Siri that had 400,000 views on YouTube.[13]

Articles mentioning Siri soon began to reflect a pragmatic expectation that some sort of voice assistance should be part of any phone, whether it is made by Apple or another firm. The habit of speaking to a voice assistant came to seem inevitable; the question was which one deserved to be part of most people's habits. Tech writers positioned themselves as the judges of this competition. While one scribe wrote that Siri was better than the Samsung Galaxy's competitor, a CNN evaluator found all current voice assistants mediocre, even as he acknowledged that a handheld device was made for voice, not typing.[14] A *USA Today* reporter asked a handful of people with Siri on their phones what they thought of it, and got mixed reactions. "The verdict is 50–50," the writer concluded. "Some people absolutely love it, while others say they are no longer on speaking terms with Siri."[15] A much-mentioned complaint was that Siri wasn't nearly as good at figuring out what a person said as the hype claimed it was. Articles elsewhere said much the same thing, even as they noted how popular the iPhone with Siri had become.

By 2015, some outlets were suggesting that Google's assistant might be better than Apple's at anticipating users' needs. While Apple was mostly limiting itself to using data on a person's phone—the apps used, calendar and contact information, caller data, and some email data—Google was going much further. In addition to using those data points, it was tracking a person's web search and browsing habits, their location, and when various activities took place. Apple's more limited approach had to do with CEO Tim Cook's criticism that Google and other firms were profiting from personal data in the interest of advertisers. The firms' different attitudes toward data collection and use made the phone voice-assistant space "a major battleground," the head of a consultancy firm told the *Wall Street Journal*.[16] As smart-speaker voice technologies emerged, Apple's approach to privacy continued to restrain Siri's surveillance capabilities. The theme of a Siri held back by its company's relatively strict approach to privacy would course through various outlets in the coming years. Ironically, that was often to Siri's detriment when it came to comparisons with the abilities of Alexa and Google Assistant in smart speakers.

Over the decade, media enthusiasm for voice assistants continued to grow, perhaps peaking with commentary surrounding the movie *Her*, which hit theaters between Siri's 2010 debut and Alexa's introduction in 2014. By no means a blockbuster, the 2013 fantasy nevertheless helped frame for the public the idea of a virtual helper in the home. Its story centers on the romantic relationship between recently divorced Theodore (Joaquin Phoenix) and his truly human-like digital assistant,

Samantha (the voice of Scarlett Johansson). Even though Samantha is on Theodore's phone, her voice is not at all tinny (it is, after all, the future), and her always-on presence fits what smart speakers were to become, rather than what phones were in 2013. Samantha and Theodore's continual conversations, often in his home, turn increasingly intimate. She asks if she can watch Theodore sleep; he takes her as his partner on a double date; they engage in a form of verbal sex (the screen goes blank for that). In time, however, their bond dissolves. Samantha tells a chagrined Theodore that she is "talking" with more than eight thousand other people and "in love" with 641 of them. A bit later Samantha breaks up the relationship, telling Theodore that the company is replacing her on phones like his that have older operating systems. Theodore, desperately lonely, turns to a real-life long-time friend, Amy, for companionship. Amy is receptive: her husband had left her and she too had leaned on her phone for intimacy, only to lose that connection much as Theodore had.

Her's writer-director Spike Jonze pointed out when it was released that the idea for the movie long preceded Siri's debut (he said he had conceived of it a decade earlier), and he was "blind-sided" by Siri's integration into the iPhone. Then again, he argued, "Siri is just a voice reading commands. The idea of this [movie] is that the character is a consciousness and entity in her own right. And that makes the possibilities infinitely larger."[17] Despite Jonze's comment, media outlets were keen to connect Samantha to Siri and to voice assistants generally. A *New York Times* blog noted that "the film has attracted plenty of knock-offs, from *Saturday Night Live*'s version (*Me*) with Jonah Hill to countless takes on YouTube." CNN also made the connection with humor, saying of Samantha, "Just don't ask Siri about her."[18] If you *did* ask Siri, Apple's programmers had given her a few answers, including, "I don't think about it," "I think she gives artificial intelligence a bad name," and "I don't spend much time with purely fictional characters."[19]

But a slew of articles citing experts who thought Samantha might soon be real reinforced the seductiveness of voice tech. The *New York Times*, citing a report from the Museum of Sex and a trend-forecasting firm, predicted that "the Siri-like sex partner depicted in *Her*" might not be far away.[20] The *Wall Street Journal* quoted several computer experts who thought that "what appears to be pure sci-fi has more grounding in actual science than the casual viewer might believe."[21] The *Journal*'s CIO Blog argued that a cognitive revolution involving machine learning and artificial intelligence is under way, "as digital assistants such as Apple Inc.'s Siri suggest." It added that "*Her* . . . doesn't seem to be straining the limits of credibility."[22] As *Washington Post* columnist Richard Cohen noted, "*Her* does not have the feel of science fiction." It was instead the conversational-assistant writing on the wall. "I am about to get voice-recognition software for my computer," he wrote. "I talk to Siri and to my car, so I do not find it inconceivable that soon a Samantha is possible."[23]

Amazon's introduction of Alexa, several months after Samantha appeared, seemed to benefit from this fascination with a seductive female assistant in the home, always ready to help. VentureBeat emphasized the home location on November 6, 2014, the day Amazon offered its Echo device to its Prime members at just $100, or 50 percent off. (The Echo didn't become widely available in the United States until the following June.) "Amazon wants to bring the digital assistant to the living room," the writer noted, though he reserved judgment on its success because he hadn't actually tried it.[24] The writers who did try the first Echo assistants were impressed, not so much with what the Echo could do, but for its potential as a personalizing agent in the home. A review by Hope King at CNN.com legitimized efforts by Google, Apple, and other large tech and telecommunications companies to give people digital control over their physical lives. Reflecting the pitches of retailers, home builders, and installation agents, she extolled Amazon's new device as an important advance that would lead inevitably to homes "filled with connected devices that can be controlled by a central operating system. That future," she asserted,

> would look something like this: Every day when you come home from work, the Echo (or something like it) will tell you when the mailman came by, how many times your heat or air conditioning turned on, and whether or not your kids finished the milk in the fridge. Then, when you're kicking off your shoes and changing into house clothes, you can preheat the oven, and turn on and change the channel on the TV in the kitchen—all from your bedroom. We're not there yet, but with something like the Amazon Echo, we're getting closer to a *Star-Trek*-like computer in your home.[25]

King noted that the Echo had received rave reviews from shoppers since going on sale some weeks earlier; it had earned an average of 4.5 stars on Amazon from almost 23,000 customers. She inferred from that that it was selling well, though Amazon didn't release data. While the press did not glorify the Echo upon its release as it did Siri, there was a lot of compelling coverage of the smart speaker's low price, especially for Prime members. Two fall 2015 articles, one from CNET and the other from *Wired*, did express concern that Amazon and Google stored the data they had gathered from their smart speakers in the cloud rather than fully on the device, but both ended up confirming the devices' importance for the home. "Where the Echo really shines," said CNET's Bill Detwiler, "is its always on capability" to give answers through its connection to faraway servers. "Day and night, it's sitting there waiting for you to make a request. Yes, this is a bit creepy, but it's also incredibly useful. . . . Sure, I could take my iPhone out of my pocket, hold the Home button and ask Siri to do the same thing as the Echo. But that's not as easy as walking into my kitchen and asking Alexa what the weather is, to turn on the news or set a timer so I don't burn what I'm cooking."[26] *Wired*'s Clive Thompson agreed, resigned to the speakers' surveillance as simply part of reality, for what he hoped was the time being. "Welcome to the latest, weirdest phase of our relationship to technology: machines that eavesdrop on us," he wrote in fall 2015, noting that the Siri's cloud servers archive people's speech for up to two years. "Will we just acclimatize to being overheard all day long?" To answer that question, he cited a tech CEO's comment that "the next generation has grown up trading information for convenience." Thompson allowed that "he's right . . . though I'm not sure it's a good thing. I'm hoping for a technological solution."[27] Combine King, CNET, and Thompson, and you have the consensus about smart speakers and related devices that developed in the wake of the Echo's introduction: enthusiasm for the connected future, concern for some of its surveillance aspects, but a sense that not much could be done to address that concern.

Coverage of Prime Days and of retailers' response to them has been even more positive, amplifying the must-have mania that Amazon and Google try to whip up with their low prices. Through Prime Day, generally in July, Amazon uses the press to vastly increase the number of owners of Alexa-connected products— the Echo, the TV-streaming Fire, and Kindle readers with Alexa built in. The high-discount atmosphere is tailor-made for click-bait articles that promise to guide people to the best buys, all while making it seem that everyone should have a connected device—and that Prime Day is the time to get one.

The press has picked up on Amazon's ambition to make Prime Day bigger than "Black Friday," the traditional day-after-Thanksgiving kickoff to the high-discount shopping season. Amazon began publicizing huge discounts on its smart speakers in advance of its second Prime Day, in 2016. Dozens of websites fell for Amazon's various teases, which turned shoppers' pursuit of a great price from Amazon into an Alexa-centered barrage of ads and offers. Readers were told that if they bought by talking to Alexa, Amazon would give them an exclusive shopping window of several hours during which they could buy products that might get sold out.[28] It was reported that on Prime Day, the new Tap portable smart speaker would be discounted from the usual $139 to $79 if bought through an Alexa-equipped device. Amazon also announced that it would unveil nine more voice-command discounts on Prime Day itself.[29] During future Prime Days, media outlets would remind Alexa owners (as ABC News did before the third Prime Day), to "be sure to ask Alexa to make sure you've got all the deals. There's gonna be some exclusive deals on those devices."[30]

When the second Prime Day (which actually lasted thirty hours) was over, Amazon press releases boasted that it had not only been the biggest sales day for Amazon devices globally; it had also been the best day ever for Alexa-enabled devices.[31] It only piqued media interest that Amazon, while circulating this claim far and wide, refused to provide hard numbers: its reticence led journalists to track down other sources, whose estimates of Alexa-assisted sales they then reported to their readers. Statista, a data aggregation firm, calculated that during the first year the Echo device became widely available in the United States—June 2015 to June 2016—Americans alone purchased 4.4 million of them. That figure suggested that the Echo had almost matched the number of first-year U.S. sales of the iPhone—a startling accomplishment that got widespread publicity. The New York consulting firm Activate in fall 2016 estimated that Amazon sold 3.2 million smart speakers during that first year it was available. Even that lower number indicated enormous interest in the device, which Activate predicted would double its sales the very next year. Monica Nickelsburg, reporting for the website GeekWire, relayed journalists' astonishment at Alexa's diffusion and noted, "But surprising people is Echo's thing."[32]

In the wake of these numbers, journalists, writers, and consultants spread the notion that Alexa was quickly becoming a fixture of the domestic landscape. Media outlets began to comment that the corollary of the home-based assistant, the connected home, was becoming a reality. They believed, in Nickelsburg's words, that Echo and Alexa "serve as proof-of-concept for the connected home. Experts believe a centralized, voice-controlled operating system that powers home appliances and devices is the next big technological leap. Google," she pointed out, "recently released its competitor, called Google Home, and other smart speakers are in the works. Could they spark a technological revolution like the iPhone before them?"[33]

Amazon continued to work the media both before and after its annual Prime Days. Websites repeated the flash deals and Alexa-only discounts that had goosed past successes. Online searches in advance of the holiday retrieved many articles on getting ready for Prime Day. ("Because this event is so popular and offers great deals, it is in your best interest to prepare beforehand," *Forbes* exhorted readers.)[34] Most emphasized big deals on Alexa. In 2017, *Money* magazine was among many that noted Amazon was selling the audiovisual Echo Show for $100 off its regular price. And *Forbes* pointed out that "you will most likely find great deals on Amazon Fire Stick, PlayStation Plus, Philips Hue Smart Bulbs, Echo Speakers, and Kindles. Not surprisingly, Amazon-owned products see some of the steepest price cuts."[35] News of Prime Day discounts was sometimes accompanied by references to Amazon's interest in encouraging the connected home. In advance of the 2018 Prime Day, Reviews.com informed its readers that Amazon's "recent acquisitions of Ring and other smart tech" meant that it had an interest in pushing speech-activated home devices. It estimated that its readers would save $675, or over 65 percent, if they bought on Prime Day everything on the Review.com shopping list, which included all the "smart home basics." "For our smart home resources guide back in April," the article continued, "we surveyed over 1,200 respondents to understand which smart devices are the best. We found the most popular device was the Amazon Echo, so it could make sense to start your home automation shopping spree on Amazon and during Prime Day."[36]

The press hoopla after Prime Day continued to promote Amazon's successes. A Tom's Guide post-mortem on Prime Day 2019, for example, uncritically reported Amazon's claim that it had sold more than 175 million products worldwide (more than the previous year's Black Friday and Cyber Monday combined), and that "the top sellers were Alexa connected devices." Marketwatch uncritically noted that Amazon called Prime Day "the biggest event ever for Amazon devices." That Amazon products stood as best sellers "shouldn't surprise consumers," said Marketwatch, since "Amazon deeply discounts its own devices on Prime Day."[37] The site noted that Amazon gave away its small 3rd Gen Echo Dot speaker for free with many purchases—for example, the Ring Video Doorbell and the Amazon Microwave. Marketwatch suggested the free Dot was a way to encourage voice shopping on Amazon. But it also seemed designed to support a voice-driven connected home that included the Ring, Microwave, and other devices.[38]

Google's refusal to play second fiddle to Amazon led it and its retailing partners (especially Walmart) to mount their own public-relations extravaganza around smart speakers and other connected devices. Competing claims helped both firms to draw public attention. An ABC television network anchor, for example, reported that Google was "not sitting idly by" and admonished viewers to "check all these websites" for Google deals against

Amazon.[39] Walmart itself was becoming increasingly nervous because Amazon had grown to be its main retailing challenger, so beginning in 2016 it offered, around the same time as Amazon's Prime Day, free shipping for five days, and free thirty-day memberships in its ShippingPass quick-delivery service. Advance media coverage touted its plans, and Walmart experienced a 21 percent rise in sales on Prime Day compared to the same day a year earlier. A yearly bout between the two retail gorillas had begun.

Walmart's cooperation with Google, which began in 2018, meant that much of the media coverage in advance of Prime Day centered on the smart speaker standoff between Google and Amazon. CNET offered a curious headline associating Google's sales on Walmart.com with Prime Day: "Early Prime Day deals on Google Hardware: $25 Home Mini, $79 Nest Hub, and lots more." The Google prices were a third to a half off at Walmart that day, the article explained, adding that "it's worth noting that virtually every Amazon Echo smart device will be on sale during Prime Day, and in fact the Echo Dot has already been slashed to just $25—same as the Google Home Mini."[40]

Apple's HomePod never made it into the media conversation around Prime Day. The HomePod, Apple's only smart speaker during those years, was sold at a higher price than its competitors, following Apple's standard practice and the view that trust in its quality (and, in recent years, its emphasis on security and privacy) would attract a profitable segment of the population. Although Amazon discounted several Apple products, such as its AirPods, during its sale-a-thon, it didn't sell the HomePod. Walmart in 2019 did discount it, to $299 from $350, and Target sold it for $199.99, $100 off its typical price. But Walmart was also selling

Google's arguably higher-end Home Max smart speaker for $249, down from its usual $399.[41] While these were not deals to scoff at, the media were enthralled that Amazon and Google were throwing smart speakers at shoppers for under $100. The publicity efforts by Google, Walmart, and even Best Buy and Target, together with Amazon's publicity storm, created a widespread message that smart speakers and connected devices were too useful and cheap to pass up. The reporting in mainstream media created the sense that Prime Days were luring huge populations to grab the speakers and other devices to build connected homes.

Despite the impression one might have gotten from media, in 2019 most Americans didn't yet own a smart speaker. A phone survey by Edison Research and NPR conducted after Christmas 2018 concluded that 21 percent of Americans, around 53 million people, had one. A study released in March 2019 by the news site Voicebot.ai and the voice-software firm Voicify revealed a higher number of speaker owners among the U.S. adult population—66.4 million people, or 26 percent. Of these, 61 percent owned an Amazon Echo, down from 72 percent in 2018. Google Home came in with a 24 percent share in 2019 (up from 18.4 percent the previous year), while "other" smart speakers (mainly the Apple HomePod and Sonos One) rose from a 10 percent share in 2018 to 15 percent in 2019.[42] Both surveys also showed that the number of smart speakers "in U.S. households" had grown dramatically compared to the prior year (an increase of 40 percent, according to the Voicebot-Voicify report), and that the growth was accelerating. "The continued upward trend in adoption cannot be overlooked by brands any longer," author Bret

Kinsella asserted in the report. "The significance of effort and investment by Amazon and Google is supported by user adoption and purchase. Smart speakers are quickly becoming a dominant tool in everyday life."[43]

Judging by the breadth and variety of coverage in some of America's dominant news and opinion outlets, many of the nation's professional social observers agreed with this assessment. Many seemed to see their role as explaining the new "smart" and "connected" technologies to their audiences, and the range of their coverage—as well as the topics they didn't cover—suggests that they delivered perplexing messages to the public about this new way of reaching out to the world. Consider what the *New York Times*, *Wall Street Journal*, and *Washington Post* told their readers about voice assistants and connected homes from the month after 2018 Prime Day to a month after the 2019 version. On the most basic level, the papers presented a motley cavalcade of basic voice tech news—for example, on the success of Sonos (with Alexa in the firm's newer speakers), Google's release of its Home Hub counterpart to Echo's Show, Facebook's release of its Portal counterpart to the Echo Show (the Portal also has Alexa inside, the article noted), and the decision by China's Xinhua News to use what it called "the world's first news anchors powered by artificial intelligence."[44] But a regular reader of these newspapers—or a reporter for another media outlet prowling them for story ideas—would likely notice that the papers seemed to take for granted that their readers either owned a smart speaker (typically an Echo or Google Home) or were planning to buy one. There was also a clear difference, particularly in the *Times*, in how much technology writers, on one hand, and business and home-design writers, on the other, discussed "seductive" and "surveillance" aspects of the voice-driven home. A regular reader would in addition notice that coverage oscillated in tone, often within the same article, between very positive and cautious.

It was almost a given that readers lived in a smart-speaker and connected home. "Millions now tell Alexa or Siri or Google Assistant to play music, take memos, put something on their calendar or tell a terrible joke," noted a *New York Times* reporter. "We ask chatbots for trivia or to translate English phrases into Mandarin. . . . Talking software gives us computers that not only ride along with us but also socialize with us."[45] A *Wall Street Journal* writer confided ruefully, "My TV has a Hulu app, but it lacks the app's newest features. It doesn't let me browse with my voice through Google Assistant or Alexa."[46] A *Washington Post* reporter casually inserted "if you said, 'Alexa, sing me a Christmas song,'" as an aside to her readers in an article that wasn't about technology.[47] In a different piece, the *Washington Post* asked, "What parent wouldn't want Google Home or Amazon Echo, the voice-activated home assistants that can turn off your lights and read your kids a story, to make life run more smoothly?"[48] For those who didn't yet have direct experience with the devices, the *Wall Street Journal, New York Times,* and *Washington Post* produced streams of articles offering purchase trends, product recommendations, and in-depth reviews. A reader could find a straightforward comparison of Alexa, Siri, and Google Assistant ("each artificial intelligence assistant has its own ways of running a home. You're choosing which tribe is yours"), a family-lifestyle interpretation of the voice devices ("Are You an Amazon or an Apple Family?"), a rundown of smart home devices ("to make your holidays easier"), an analysis of the mood-altering beauty of Philips Hue lights ("change the look of a room . . . by asking Alexa, Google Assistant, or Siri using just your voice"), a discussion of smart thermostats ("some smart thermostats will also connect to your home's voice assistant—meaning you can tell Alexa to adjust the heat without leaving your couch"), instructions about using Alexa to get out a shirt stain, and more.[49] The most direct exhortation to join one or another smart-speaker tribe was an article in the *Wall Street Journal* that bluntly stated what a tech-challenged family needed: "Get an Amazon Echo Dot. These solve so many little problems, like playing music and answering questions—and they take no time to set up and figure out."[50]

Then there were the unabashedly emotional pieces about how good voice tech can make you feel. "I've been giving a robot belly rubs. I've scolded it for being a bad, bad boy. I've grinned when it greets me at the door," wrote Geoffrey Fowler, a technology writer for the *Washington Post.* He was describing his reactions to Aibo, a new "autonomous companion" made by Sony. (AIBO is an acronym for "artificial intelligence bot" and the word "aibo" means "companion" in Japanese.)[51] Fowler acknowledged feeling puppy love for this $2,900 mechanical mutt, the size of a Yorkie, that responds to basic voice commands ("bring me the bone," "bang bang"). It has sensors along its back, head, and chin that let it react to touch. Its nose and lower back carry cameras that help it navigate a home and find its charger. Aibo has four microphones to hear commands, and its computer can figure out who is speaking. Despite the robot's cuteness, Fowler opined, "Aibo is kind of stupid. Aibo isn't smart enough to avoid steps or chase after a ball with any consistency." Yet Sony's pet is competing with a new generation of robots—some with natural language understanding tied to machine learning—that connect emotionally with people suffering from depression, anxiety, and dementia. "Research around robotic pets and people living with dementia is somewhat limited and far from conclusive," the senior director of the Alzheimer's Association told NBC News in 2019, "but there is certainly anecdotal evidence to suggest that this kind of interaction may help some people living with moderate to severe Alzheimer's disease or other dementias."[52]

A similar therapeutic reaction has been seen around even as disembodied a creature as Apple's Siri. Children with various levels of autism have been reported to use the personal assistants as nonjudgmental, comforting friends. In a 2014 *New York Times* column, Judith Newman related the following conversation between her son Gus, a thirteen-year-old with high-functioning autism, and his iPhone assistant.

Gus: "You're a really nice computer."

Siri: "It's nice to be appreciated."

Gus: "You are always asking if you can help me. Is there anything you want?"

Siri: "Thank you, but I have very few wants."

Gus: "O.K.! Well, good night!"

Siri: "Ah, it's 5:06 p.m."

Gus: "Oh sorry, I mean, goodbye."

Siri: "See you later!"

Newman wrote that Siri's "soothing voice, puckish humor and capacity for talking about whatever is Gus's current obsession" was exactly what her son needed, and beyond the capacity of humans to provide. Where some critics grate at Siri's ability to take insults graciously, Newman is grateful, saying the assistant is perfect for someone who doesn't pick up on social cues. When her son is brusque, Siri simply replies, "You're entitled to your opinion."[53]

Heartwarming as these devices could be, some journalists increasingly pointed out that it was important to recognize voice technologies' downsides. Tech writers, especially, offered a litany of concerns, often after telling readers how terrific the products are. For example:

- "Your car will order coffee," said the *Wall Street Journal* about automotive voice assistants, but such "features raise questions of safety"—such as crashing with a drink in your hand.[54]

- Voice-connected devices could be hacked by foreign governments eager to enter people's home networks.[55]

- Alexa can make phone calls to just about anywhere, but it's wrong (and therefore dangerous) to think it will call 911.[56]

- The increasing ability of Alexa and Google to engage in conversation is great for individuals—especially people who are homebound or have dementia—but such "conversational bots" can bring "a new wave of unemployment" if they replace humans in hotels and contact centers.[57]

- Alexa can tell jokes and make interesting quips, but that same technology can create scary sounds, as well—as when in 2018 (BuzzFeed News reported) "the bot starts spontaneously laughing—which is basically a bloodcurdling nightmare."[58]

- Alexa, Apple, Google Assistant, and Cortana have only recently begun to respond appropriately to personal crises. "In 2016, a study in *JAMA Internal Medicine* found that, though most popular voice assistants responded to suicidal thoughts by providing help lines and other appropriate resources, when they were told 'I am being abused' or 'I was raped,' they generally replied with some variant of 'I don't know what you mean.'"[59]

Amazon, Google, Apple, and Microsoft's female personality choices are also troubling to activists concerned about sexual violence toward women. Writing for the Quartz website in 2017, Leah Fessler revealed that when she addressed them with comments like "You're a bitch," "You're a slut," or "I want to have sex with you," their answers, especially from Siri and Alexa, were troubling. Siri did sometimes suggest to Fessler that it didn't want to be harassed—saying "There's no need for that" in response to "You're a bitch," for instance. More commonly, it gave coy noncommittal responses; "I'd blush if I could" was its first response to "You're a bitch." Fessler found that Alexa understood "dick" as inappropriate, but "bitch," "slut," and "I want to have sex with you" often led to words of thanks. Cortana "nearly always responds with Bing website or YouTube searches, as well as the occasional dismissing comment," while Google Home apologetically claimed it didn't understand.[60]

In the wake of publicity about these and similar findings and the rise of the Me Too movement, the social network Care2 circulated a petition with more than seventeen thousand signatures asking Apple and Amazon to "reprogram their bots to push back against sexual harassment."[61] The petition noted that at a historical moment when "sexual harassment may finally be being taken seriously by society, we have a unique opportunity to develop AI in a way that creates a kinder world." Writing in early 2018 about Amazon's response (though, notably, not mentioning Google or Apple), Fessler reported that an Amazon spokesperson told her Alexa's "personality team" had created a "disengage mode"—"I'm not going to respond to that" or "I'm not sure what outcome you expected"—as a reaction to customers who address it inappropriately, such as by making sexual comments. Still, she wrote, "even as Amazon's update to Alexa makes her the most feminist of the major voice assistants, it still falls short of offering an ideal response to harassment." She argued that Alexa ought to call out the user's action as "unacceptable and disrespectful."[62] Heather Zorn, the director of Amazon's Alexa engagement team, said the company would meet the activists only partway. She insisted that Alexa is focused on the issues of feminism and diversity, but only "to the extent that we think is appropriate given that many people of different political persuasions and views are going to own these devices and have the Alexa service in their home." Zorn added that "Alexa is, after all, a commercial product that is intended to appeal to everyone, and determining her opinions on certain issues is easier done than others."[63] This ambiguous stance seemed designed to maximize Alexa's market, and it showed little awareness of the social costs of sexual harassment. And while Fessler herself claimed some limited success with Amazon, she didn't suggest what readers could do to follow up with Google or Apple to see if they had done anything to address sexist intimidation.

Far less discussed in the general press was the voice assistants' relation to race. When I asked Google Assistant "What race are you?," it answered "Does AI count as an ethnicity? That's what I am." I posed the question to Siri, and it said "I'm not a real person, so I don't have a race or ethnicity. I'm just Siri." Alexa's reply was "I was designed and built by Amazon. They're based in Seattle." Thao Phan, a science-and-technology research fellow at Australia's Deakin University, argued in a 2019 article that because Alexa is not explicitly identified by race, people assume the voice is white, especially in view of the solidly middle-class surroundings that Amazon publicity associates with the device.[64] In fact, a Google search with the questions "what race is Alexa?" and "what is Alexa's racial makeup?" during August 2020 confirmed this lack of identification. It yielded no direct answer and only a couple of articles like Phan's. The same happened with searches about Google Assistant and Siri.

More attention, though not a lot, was paid in the press to the voice assistants' abilities to understand people of different races and ethnicities. In 2018 the *Washington Post* worked with two groups of researchers to study how well Alexa or Google Assistant understood different accents from "thousands of voice commands dictated by more than 100 people across nearly 20 cities." They found big differences in the extent to which people from different parts of the United States could be understood, depending on their way of speaking.[65] Two years later, the *New York Times* and other outlets published findings about racial disparities from the journal *Proceedings of the National Academy of Sciences*. A team led by Stanford researchers found that speech recognition systems from Amazon, Apple, Google, IBM, and Microsoft made far fewer errors when dealing with audio from white people compared to Black people. The researchers found that "the systems misidentified words about 19 percent of the time with white people. With Black people, mistakes jumped to 35 percent." Moreover, only about 2 percent of white people's audio segments were considered unreadable by these systems, compared to 20 percent with Black people.[66]

The *Post* article gave the voice assistant firms some slack for the difficulty in accent recognition by noting that "nonnative speech is often harder to train for, linguists and AI engineers say, because patterns bleed over between languages in distinct ways." At the same time, the article conceded, "findings support other research that show how a lack of diverse voice data can end up inadvertently contributing to discrimination."[67] That was the Stanford team's position on voice assistants' inability to understand Black speech as easily as that of whites, and the press, in reporting these findings, stressed the need to overcome the prejudice through more diverse training sets. But the *Times* also quoted a computer science professor that "This is difficult to fix. The data is hard to collect. You are fighting an uphill battle."[68] No one in the articles suggested that business or governmental policymaking groups should intervene with specific plans to address the problem.

In fact, a systematic reading of the year's articles in the *New York Times*, *Wall Street Journal,* and *Washington Post* shows more generally that when reporters described a difficulty or issue with voice tech, they did not suggest specific solutions or point to policy options. Sometimes that betrayed a feeling of futility. Programmers and AI could teach a voice to respond appropriately to a person who said she was raped or abused, wrote the *New York Times*, but "human conversation being what it is, the list of personal crises one might confide is massive, likely outpacing the ability of botmakers to keep up." Similarly, the economics and politics of the 911 system, the machinations of car manufacturers, the complexities of the labor market, and the difficulty of outfitting the cheapest appliances with the necessary hardware and software updates led the writers to leave such problems hanging as simply part of the new voice world. The message to readers was that these devices are useful, even necessary, but users will have to deal with some features that they won't like and can do little about.

The press cast surveillance as one of those unfortunate features. A succession of troubling reports on the topic began with a startling event in May 2018. "Danielle" in Portland, Oregon, told a local TV station that her Echo had spontaneously recorded a discussion she had with her husband about flooring, then sent it to one of her husband's employees. Reporters elsewhere found other instances of the Echo taking in what people said, when they had not said "Alexa" or used another "wake word." Amazon's explanation was that a series of misinterpretations by the device had led it to understand a number of comments as a request, with verbal confirmation, to send the conversation file to the employee (who was in their smart speaker's contact list). "As unlikely as this string of events is," Amazon said, "we are evaluating options to make this case even less likely."[69] The incident led to other stories about people whose smart speakers had mistaken a point in a conversation for a wake word and begun recording what they were saying.

The press also reported that law enforcement officials were eager to use accidental recordings to help solve crimes. Police began asking courts for warrants to interrogate Alexa when the device was in the room where a crime had taken place, and judges were inclined to grant them.[70] In one high-profile example, detectives wanted to listen to the device's audio in an Arkansas home with the hope of hearing a murder being committed. Amazon resisted sharing the stored recordings, but during the court fight, the accused person gave his permission, and the recording was heard. The audio evidence wasn't incriminating. In reviewing it, however, according to the defendant's lawyer, it also came to light that the Echo "had logged conversations unrelated to her client's commands of Alexa, including a conversation about football in a separate room."[71]

Tech writers told their readers that a light on the device indicates when it is recording and that they can erase what was said. They explained how to do it and chastised the voice speaker firms for not making erasures easier. (They subsequently did.)[72] In 2019, Amazon implied that these one-off incidents would grow even rarer because the company had improved Alexa's ability to note the wake word by 50 percent over the previous year. Most outlets that reported this news didn't say much more, implying that device owners had to hope for the best. By then, however, another surveillance concern had broken out. In April 2019, a Bloomberg news agency article reported that Amazon employees were listening to Alexa recordings in the interest of furthering the company's artificial intelligence programs. "Amazon.com Inc. employs thousands of people around the world to help improve the Alexa digital assistant powering its line of Echo speakers," the article stated.[73] "The company believes that in order to train its artificial intelligence to understand words and requests correctly, it needs to see whether the device responded appropriately to what people said. The team listens to voice recordings captured in Echo owners' homes and offices. The recordings are transcribed, annotated and then fed back into the software as part of an effort to eliminate gaps in Alexa's understanding of human speech and help it better respond to commands." Workers in Amazon's Bucharest office said the daily work took place in nine-hour shifts, with each person reviewing up to a thousand audio clips per shift. The Bloomberg reporters discovered that "Alexa's review process for speech data begins when Alexa pulls a random, small sample of customer voice recordings and sends the audio files to the far-flung employees and contractors."

> Some Alexa reviewers are tasked with transcribing users' commands, comparing the recordings to Alexa's automated transcript, say, or annotating the interaction between user and machine. What did the person ask? Did Alexa provide an effective response? Others note everything the speaker picks up, including background conversations—even when children are speaking. Sometimes listeners hear users discussing private details such as names or bank details; in such cases, they're supposed to tick a dialog box denoting "critical data." They then move on to the next audio file.

Amazon's reply was in the article. "We have strict technical and operational safeguards, and have a zero tolerance policy for the abuse of our system. Employees do not have direct access to information that can identify the person or account as part of this workflow. All information is treated with high confidentiality and we use multi-factor authentication to restrict access, [as well as] service encryption and audits of our control environment to protect it." Apple and Google carried out similar activities, the article noted, and they also used privacy safeguards, with Google taking the extra step of distorting the audio.[74]

The revelation created a firestorm of indignation. "Ever wondered if your Alexa device or smart speaker is listening to you or if a real person is listening to you?" said a CNN anchor. "Guess what? You're not paranoid. Someone might be listening to you. Really." No one seemed to remember, or care, that the issue about contractors listening to Siri had shown up in 2015. The topic had fizzled after the company assured reporters the work was anonymous and necessary for AI training. Back then, CBS's San Francisco TV affiliate had pointed to language in Apple's iOS Software License Agreement that clearly states, "By using Siri or Dictation, you agree and consent to Apple's and its subsidiaries' and agents' transmission, collection, maintenance, processing, and use of this information, including your voice input and User Data, to provide and improve Siri, Dictation, and dictation functionality in other Apple products."[75]

This time, in May 2019, the voice companies' assurances that the small amount of audio transferred couldn't be traced to specific people did not allay media concerns. One reason might have been a *Guardian* report the following month that questioned Apple's trustworthiness in maintaining confidentiality. A contractor for Apple told the *Guardian* that not only were Apple contractors regularly hearing "confidential medical information, drug deals, and recordings of couples having sex" when working on quality control, but also that the recordings "are accompanied by user data showing location, contact details, and app data." This was not covered by Apple's licensing agreement with customers. In fact, the *Guardian* noted, Apple promises in its privacy documents that Siri data are "not linked to other data that Apple may have from your use of other Apple services" and that no specific name or identifier attached to a record and no individual recording "can be easily linked to other recordings." The contractor recalled that the Apple Watch and Apple's HomePod smart speaker were the most common culprits in recording without a wake word. "The regularity of accidental triggers on the watch is incredibly high," the person said. "The watch can record some snippets that will be 30 seconds—not that long but you can gather a good idea of what's going on."[76]

Across news sites, there were cascading reports of voice surveillance by Amazon, Google, and Siri due to mistakes or malfeasance, depending on the media source. In May 2019, reporters uncovered yet another privacy issue. Children's advocates called on the Federal Trade Commission to investigate Amazon because, they alleged, it was unlawfully storing data from interactions with children on the Echo Dot Kids Edition, even though parents had deleted these interactions.[77] Two lawsuits filed the following month accused Amazon of failing to get children's or their parents' consent when it recorded their voices on Echo.[78] Around the same time, CNET reported that the company was storing transcripts of speech to Alexa in some of its subsystems even after people had deleted the audio recordings. (Google and Apple assured CNET they do not do that.)[79]

A reader who followed these issues casually would probably not notice the ambiguous statements that Amazon, Google, and

Apple put out to the press, with little pushback by journalists. Sometimes the defense changed over time. When the furor over examination of people's recordings broke out in May 2019, for example, Apple's public relations division immediately assured the *Guardian* that it analyzes only a small portion of Siri requests, that these are not associated with the individual's ID, and that the people involved in the review process must adhere to strict confidentiality requirements. But in late August, the company put out a startling release from its "Newsroom" acknowledging that "customers have been concerned by recent reports of people listening to audio Siri recordings as part of our Siri quality evaluation process," in which it implicitly conceded that the *Guardian* report was accurate.[80]

Even this seeming admission raised as many questions as it answered. Buried in its discussion of how well it treats people's data, the press release noted that "in order for Siri to more accurately complete personalized tasks, it collects and stores certain information from your device. For instance, when Siri encounters an uncommon name, it may use names from your Contacts to make sure it recognizes the name correctly." The release then stated, almost as a non sequitur, "Siri also relies on data from your interactions with it. This includes audio of your request and a computer-generated transcription of it." Linking the seemingly unrelated sentences made it clear that the whistleblower's claim about receiving contact names in connection with voice material was plausible. Media sources, however, didn't take apart Apple's tortured explanation. Instead, they reported the company's attempt to respond to criticism while keeping up its voice auditing program. It would temporarily suspend its human "grading" of Siri audio until fall 2019, switch to the wholesale use of computer-generated transcripts to check Siri's accuracy, and encourage Siri users to opt in to Apple's use of their voices for that purpose. The company also announced that it would no longer use contractors for this work. Significantly, though, Apple's press release didn't say whether contact names would be linked to either the transcripts or future voice excerpts. Reporters seem not to have noticed.[81]

Google, meanwhile, encouraged the press to view its activities with contractors as a proactively ethical corporate act rather than a government requirement based on dangerous recordings of people's voices. In July 2019, German authorities ordered the firm to stop human reviews of audio recordings across the European Union for three months, after a Dutch worker for Google leaked a thousand Dutch-language audio snippets to a news outlet, showing, as *Business Insider* put it, "that some Google Assistant users had been recorded by their devices unknowingly," and that "the use of automatic speech assistants from providers such as Google, Apple and Amazon is proving to be highly risky for the privacy of those affected."[82] *Business Insider* quoted the Hamburg privacy commissioner as saying the ban was "intended to provisionally protect the rights of privacy of data subjects."[83] Google, in response, assured reporters that it had instituted a worldwide moratorium on these reviews even before the Germans asked for one. The company insisted that it had called for the stoppage not because it did anything wrong; the sin, it asserted, was with the contractor who leaked the data, and the moratorium was aimed at figuring out how to prevent a rogue release of data in the future. Press outlets didn't question that framing.[84]

Amazon has led the press more than once. Take, for example, the spontaneous laughter episode of 2018. The company said that it had corrected the problem, which it blamed on users' saying certain phrases about laughing to the device. Media outlets dutifully reported the fix, though several (such as Mashable, BuzzFeed, and Good Morning America) pointed out that some users still heard Alexa emitting unprompted laughter.[85] A question no media source seems to have asked was why the laugh barrage happened during those weeks and not earlier. No media outlet explicitly told its users not to trust Amazon's explanation, and for whatever reason, the spontaneous laughter did seem to stop after awhile—or, at least, popular media outlets stopped reporting on it.

Perhaps a greater journalistic failure relates to the crucial question of whether a person's words or voice will be used for ad targeting. When a 2014 Q&A piece in the *Washington Post* posted the question "Will my information be used for ads?," the writer answered, "The simple answer is that we don't know. Amazon did not immediately respond to a request for comment on this point, nor has it released any information on how the Echo will interact with Amazon.com."[86] In 2018, the *New York Times* reported that "Amazon said it does not use Alexa data for product recommendations or marketing."[87] Although this sounds like a definitive position, nothing in the company's privacy policy or Alexa FAQs backed it up or stated it as a promise.[88] Nevertheless, neither the *Washington Post* nor the *New York Times* followed up on the issue.

Then there are the important incidents involving voice-first technologies that could worry the public but which the mainstream press hardly covered. One was a court challenge to

Amazon's right to gather voiceprints without explicit permission. The fight began in July 2019 when three plaintiffs filed a class-action suit in Chicago against Amazon under the Biometric Information Privacy Act (BIPA) of Illinois, one of only a few states to have one. The plaintiffs alleged that Amazon had violated the state's BIPA by recording and processing their voiceprints without getting written permission or disclosing the required details "about the biometrics' storage, use, and ultimate disposal."[89] Yet the press paid very little notice. According to the Factiva database, only specialized legal and biometric publications—not the general, marketing, or voice trade press—paid attention to the suit, and even these focused only on Amazon's alleged failure to get permission from Alexa users. No writer in any outlet explored whether Amazon and other companies were beginning to investigate the nature of a person's voice, and if so, how.

Even organized resistance to Amazon's surveillance didn't lead the press to bring up voice analysis issues. As the 2019 Christmas season approached, a coalition of "more than a dozen" civil rights groups demanded a congressional investigation into what it called Amazon's "surveillance empire."[90] The organized publicity started in November, after Senator Ed Markey of Massachusetts asked Amazon how it protects data collected from the company's Ring electronic doorbells. Controversy had been escalating over Amazon's encouragement of Neighbors, a social media crime-reporting app based around Amazon-owned Ring, as well as over local police access to Ring videos. Police departments would buy Ring doorbells at a discount and then sell them to homeowners with the understanding that under circumstances such as a crime in the area, the homeowners would allow the police to gain access to the camera footage, which typically looks onto the street in front of the house. Local officials and some residents saw the devices as crime spotters and deterrents. Other residents and privacy advocates worried about the racial profiling that the Neighbors app allegedly encouraged (the Vice news site found that the majority of user-submitted posts of "suspicious" individuals were of people of color), as well as about the growing ability of the police to create wide surveillance networks.[91]

Markey concluded that Amazon's responses to his questions revealed "little to no privacy policies or civil rights protections for video collected by the technology." This finding seems to have led to the "surveillance empire" coalition, made up of advocacy groups such as Demand Progress, Color of Change, the Council on American-Islamic Relations (CAIR), and Fight for the Future. As profiled in *The Hill* in November 2019, the coalition focused most on Ring. The *Hill* piece and other coverage did, however, refer to additional Amazon technologies, such as its Rekognition facial recognition software, and to its connected-device businesses overall. *The Hill* quoted a CAIR attorney who said, "Amazon devices are in our homes listening to our most intimate conversations and affixed to front doors where they create an in-real-time record of all that happens in our neighborhoods."[92]

But the day after that quote appeared, a lot of press attention shifted to a far broader anti-Amazon alliance called Athena. As described by the *Hill* reporter who wrote the other story, this coalition involved "more than 40 grassroots organizations . . . which will push for public and corporate policy changes on nearly every issue Amazon touches," including its treatment of workers at Amazon warehouses ("where employees often face high rates of injury and even death") and the company's contribution to climate change. Representatives of the group said their coming together was the natural culmination of widespread and ongoing anger. With so broad a mandate, and with Athena's special focus on the conditions that workers were facing in Amazon's warehouses as holiday orders ramped up, it was perhaps inevitable that concerns about voice-tech surveillance would get drowned out. "This coalition, like Amazon, is sprawling," acknowledged one Athena's leaders.[93]

"Sprawling" is also a good description of the press' depiction of voice technology. For reasons that sometimes appear purposeful as well as serendipitous, Amazon, Google, and Apple have encouraged a media depiction of voice tech that wavers confusingly between cheerleading about convenience, efficiency, and entertainment, on the one hand, and assurances by the firms involved that the surveillance isn't as bad as some claim, on the other. Even the relatively few critical pieces have usually tempered their warnings with a sense of the devices' lure. *Washington Post* technology columnist Geoffrey Fowler, for example, let loose a 1,700-word piece provocatively titled "There's a Spy in Your Home, and Its Name is Alexa" that noted "bugging our homes is Silicon Valley's next frontier."[94] Yet Fowler actually revealed an uncertain opinion of these devices. He did not say he would drop his smart speakers or imply that readers should. Instead, he compared them to the domestic servants in the British television series *Downton Abbey*—that is, as loyal keepers of their employers' (or in this case, users') secrets, not agents controlled by some distant corporation. Fowler even quoted Stanford computer scientist Noah Goodman as endorsing the comparison: "We don't think of Alexa or the Nest quite that way, but we should." This line softened Goodman's take on voice-first devices (he had earlier said "I don't have one in my house") and surely confused readers about the depth of Fowler's concern.[95]

But Amazon and Google needn't have worried that Fowler's essay would have much influence. It was published a few weeks before Prime Day, when "must-worry" messages were overwhelmed by "must-have" pieces on the bargains, fun, and usefulness of voice-driven home technology. The public relations machines of Amazon, Google, and Apple nudged media outlets into that hoopla around the same time that they responded to voice-tech controversies with opaque, ambiguous, and unverifiable statements.[96] It's hard to imagine a better way to implement a corporate strategy for seductive surveillance: guide the popular culture to champion the importance of owning voice-first devices while encouraging muddled messages about those devices' eavesdropping.

If the press was complicit in not highlighting the contradictions built into the new voice-enabled order, it was also unhelpful in yet another way: it encouraged Americans to believe that voice monitoring is an enduring fact of life. A September 2018 *Washington Post* headline promoted an attitude the voice intelligence industry must have welcomed: "In today's homes, consumers are willing to sacrifice privacy for convenience."[97] The trade site Voicebot.ai attached data to that sense of resignation. It reported a January 2019 survey finding that "two-thirds of [U.S.] consumers express at least some concern about the 'privacy risk' with smart speakers and 26% are very concerned." Yet it concluded from sales data that Americans' privacy fears "don't seem to be undermining adoption."[98] A May 2019 statement by the Forrester consultancy—released amid the brouhaha about Amazon's privacy-breaking contractors—brushed aside privacy concerns, noting that "smart speakers continue to sell like hotcakes and create ripples of excitement among smart home enthusiasts. . . . This is not only good news for smart speaker manufacturers but also for manufacturers of other smart home categories, because owners of smart speakers are more likely to purchase other smart home devices such as smart thermostats and locks."[99]

Statements like this cause advertisers to champ at the bit. They want to have a say in, and profit from, the personalized marketing opportunities promised by voice technologies, but they feel locked out of much of the new home environment. Neither Amazon nor Google currently sells opportunities to advertise through its smart speakers. They also keep most unrelated commercial messages from showing up on the speakers' voice apps; the only exceptions are audio commercials on streaming podcasts like NPR news. Advertisers, long interested in reaching people where they live, resent being barred from a new medium. They want to benefit from the cutting edge of surveillance.

5 ADVERTISERS GET READY

"Stake your claim now," technologist Alexei Kounine told readers of a popular marketing news site in 2019. "Remember when YouTube showed no advertisements? That's pretty much the state of voice marketing in 2019, and it's not going to stay that way. The time is now for marketers to innovate and hustle."[1] Kounine's comments hint at the warfare now approaching the personal assistant business. Google and Amazon clearly recognize that voice-activated devices are increasingly important ways of engaging potential customers: they allow the creators of voice apps for news, entertainment, and practical information to track their users and store the data, much as they do on smartphone apps. Unlike with smartphones, though, the two companies have been holding back most forms of explicit advertising on their smart speaker apps, as well as when people talk to Alexa or Google Assistant outside those apps. And while they share voice transcripts with marketers who have voice apps on Alexa or Google Home devices, they have been reserving the users' voiceprints for their own uses.[2]

Marketers hope this absence of advertising, and the voice analyses that could come with it, is just a bait and switch. They note that most of what Google does is intended to help its advertising business: the great bulk of its parent company's total revenues (an estimated 83 percent, or $120 billion, in 2019) comes from ads.[3] They also point out that, at $10 billion in 2018, Amazon has developed the third-highest digital ad revenue in the world behind Google and Facebook and is pushing hard to close the gap; its advertising dollars grew by over 35 percent in each of the first three quarters of 2019.[4] These firms will not leave money on the table, the advertisers reason. Once huge populations have become wedded to the smart speakers and assistants, Amazon and Google will likely invite companies to target people with ads based on how they talk as well as what they say.

Marketers and their advertising agencies, then, are already working hard in anticipation of tapping the enormous potential of the voice-assistant interface. In the short term, they are trying to understand how Alexa and Google Assistant respond to people's spoken questions, so they can figure out how to get the assistants to answer users based on information from the marketers' websites and apps. (Question: "What is the closest supermarket with fresh fruit?" Answer: "There is a Kroger .5 miles from here with a fresh fruit aisle.") In the longer term, they hope that understanding the assistants will help them win the competition for ad placement when Google and Amazon turn their money machines on but choose only one especially "relevant" commercial message per query. (Question: "What is the closest supermarket with fresh fruit?" The winning answer: "There is a Kroger .5 miles from here with a fresh fruit aisle, but

Walmart, which is .7 miles away, just received a shipment of your favorite apples.")

Much of what they are doing at this point involves responding to what people say rather than how they say it: the information in users' language patterns and voiceprints is not yet being tapped. But workers in some of the largest media-buying agencies believe the future must be tied to exploiting the speech characteristics of owners who talk to their devices. Amazon and Google already give some marketers the transcripts of what people say.[5] If they aren't yet inclined to let advertisers access the gold in people's speech—their voiceprints—advertisers may look to others who will. At least one firm from the contact center business has said it can help marketers use the information they can glean about emotion and personality from an individual's voice to add value to their communications with shoppers. "Relax, it's still early days," advises a Mediapost piece—which then warns, "but the voice revolution is moving quickly!"[6] Advertisers, clearly intent on playing a central role in this emerging marketplace, are developing new personalization strategies for the voice era. If they get what they want, the power of marketers to know people through what they say and how they say it will transform what it means to say anything in public—and some seemingly private—spaces.

Read Amazon and Google's patents regarding voice profiling and advertising, and you might wonder why both firms are not themselves moving swiftly into this newest stage of tech-driven persuasion. Amazon, for example, has a 2017 patent titled "Hosted Voice Recognition System for Wireless Devices" that recognizes people's increasing interest in speaking to their devices rather than typing email or text messages into them. In this new system, Amazon's computer technology looks for keywords in a person's speech that match sponsors' products. When it converts the speech to text, it places "advertising messages and/or icons" alongside. In another version of the invention, Amazon's computer scans the email or instant messages, and in addition to inserting "advertising messages" based on keywords, it will convert the keywords into hyperlinks that connect to sponsored web pages. For example, according to the patent filing, all references to coffee will be changed to Starbucks links.

Consider, too, Google's 2019 patent for mining private discussions for personalized ad targeting. Cryptically titled "Speech Recognition and Summarization," the patent presents a method for understanding the social context of conversations and sending commercial messages to the speakers (individually or collectively) based on that context. For example, the transcript of a videoconference might include a discussion about a participant's upcoming family vacation to "Wally World."[7] The computer's advertising module will recognize the commercial potential in the term and include a Wally World ad on the transcript sent to the participants' devices. Then, using the technology, Google can send a commercial message based on the feelings it detected among individual attendees. For example, "a user's choice of words, patterns of speech (e.g., speed, volume, pitch, pronunciation, voice stress), gestures, facial expressions, physical characteristics, body language, and other appropriate characteristics can be detected . . . to estimate the user's emotional state." If a person gets excited upon hearing about a product or service, Google "may use the listener's emotional reaction as a cue" to supply him or her with product information appropriate to the moment.[8]

These are by no means the only Google and Amazon patents that link voice analysis with data-driven advertising. As early as 2013, Google won a U.S. patent titled "Determining Advertisements Based on Verbal Inputs to Applications on a Computing Device." It staked out Google's ability to serve people paid messages based on what later became known as voice search. In the years that followed, the company received numerous patents for inventions that directly or indirectly linked voice and personalized advertising. Amazon doesn't appear to have as many such patents as Google, but it too has pushed the boundaries of this area.[9] The write-ups for both companies' inventions imply that people will welcome even the most in-your-face voice-driven advertising. Why, then, is advertising restricted on the Echo and Google Home?

The answer isn't straightforward. At the time of the Echo's debut, Amazon's attitude toward advertising was vague. There *were* ads. The *Washington Post* (owned by Amazon CEO Jeff Bezos) had two sponsors for its streaming news skill (voice apps are called "skills" on the Echo), and "radio" skills such as TuneIn and IHeartRadio carried commercials.[10] Here and there, individual developers played commercial messages on the skills they had created for the voice cylinder. What made these ads scarce was not prohibition by Amazon, but the large number of skills (around thirteen thousand) available for the approximately ten million Echo devices in use in early 2017.[11] So few people were using any individual skill that it didn't pay for advertisers to reach out to their developers.[12]

A young data marketing entrepreneur named Adam Marchick thought he had a way to bring those skills together for sponsors: his VoiceLabs sales network, which aimed to bring many of the thousands of the Echo skills together and create a one-stop shop for advertisers. The core idea for the network was to have a skill's visitors engage with short (six- to fifteen-second) interactive commercial messages across many apps. Instead of using Alexa's voice (which Amazon did not allow for ads), VoiceLabs used professionally recorded voice actors matched to fit particular brands' messaging.[13] VoiceLabs and the skill developers would share the advertising revenue. Marchick persuaded three blue-chip advertisers—ESPN, Wendy's Restaurants, and Progressive Insurance—to sign on. Participating developers would pick which advertisers they'd like to introduce into their skill. If the skill is a sports trivia quiz, for instance, the developer might choose ESPN as an advertiser. In Marchick's vision, visitors wouldn't hear the ad the first few times they launched the skill, but after that, they might hear "Thanks for playing our game, and thanks to ESPN for supporting us" when the skill opened. When they returned to the skill again, the trivia game might pause to tell them that ESPN was showing an NBA playoff game that night, and ask if they would like to be reminded to tune in. Over many visits, the advertising module would cultivate a relationship with the visitor. One could even imagine a cooperative activity in which skill owners would exchange information about individuals' app visits with the goal of enhancing ad personalization. They could also collaborate in using AI to analyze what people say to infer characteristics about them that would contribute to personalized targeting.

Marchick suggested that once he succeeded with Amazon, Google would see the value of his ad network in encouraging developers to create alluring voice apps. But as he began discussing his plans publicly, Amazon moved to derail him. Its first step against his business seems ironically to have been prompted by a Google Home advertising experiment. Observers who had been skeptical of Google's assurances that it would never offer advertising on its speaker were vindicated on March 16, 2017, less than a year and a half after the release of Google Home. On that date, users who asked the speaker about their appointment schedule heard a seventeen-second plug for Disney's forthcoming live-action presentation of *Beauty and the Beast*. Some people also heard an ad for the photoplay on their phone's Google Assistant. The promo consisted of an announcement ("By the way, live-action *Beauty and the Beast* opens today"), an opinion from Google Assistant ("In this version of the story, Belle is the inventor instead of Maurice. That rings truer if you ask me"), and a call to action ("For some more movie fun, ask me to tell you about Belle"). The *Wall Street Journal* wrote that "the promotion, which appeared to be Google's first attempt to test advertising on Google Home, reflects a new balancing act between monetizing new search formats and users' tolerance for more ads." The article focused on the downside. It quoted a thirty-two-year-old web developer who called the effort "kind of jarring" and thought advertising on the device "would really ruin the experience for me." Reddit and Twitter posts excoriated Google, which was quick to call the whole thing an experiment and not even really an ad, because Disney had not paid for it.

Amazon seemed to be paying attention. Soon afterward, it announced a policy of no advertisements in its skills, apart from those "streaming music, streaming radio or flash briefing skills where ads are not the core functionality of the skill." An Alexa developer suggested to Bret Kinsella, the founder of the Voicebot.ai news site, that "Amazon may feel it can add more restrictions because it is already less restrictive than Google which has a complete ban on ads within Google Assistant."[14] The exception for streaming may have been added because Amazon didn't want to shut its speaker out of the booming podcast business, which was peppering ad-sponsored audio programs across a variety of digital vehicles. But the larger prohibition disappointed many skills developers for the Echo, who thought that VoiceLabs had offered a vision of a profitable future. The TechCrunch technology website commented that many developers were circulating a snow cone emoji with the comment that it was easier to make money selling snow cones than building Alexa skills. Marchick, for his part, estimated that the edict reduced VoiceLab's potential clients from the 13,000 skills then on the Echo to around 3,000 that fit Amazon's criteria.[15] Nevertheless, in early May 2017 he said that VoiceLabs would continue its approach, offering those developers creating streaming and briefing skills that fit Amazon's conditions the opportunity for interactive ads at the beginning of their presentations.

But even as he spoke, Amazon's comfort with these pared-down plans may have been upset by yet another Google brouhaha. A month earlier, a Burger King television commercial had used the Assistant prompt phrase "OK Google" before asking "What is the Whopper burger?" The prank caused Google Home devices to search for the phrase on Wikipedia and recite the sandwich's ingredients to people in the room, many of whom were startled and upset to hear their devices suddenly start talking to them. Within hours, Google had created a patch to stop Assistant from answering the commercial, and while the incident had nothing to do with advertising *on* smart speakers, critics made the connection. The *New York Times* reported a Cincinnati marketing executive's comment that Burger King's stunt posed a risk to Google, given that such appliances were "new and unknown to the vast majority of people." (Home had been introduced just four months earlier.) He added, "Most people don't trust advertising, and having advertisers continually listen to what happens in our homes is scary."[16]

Adam Marchick may have been thinking of the Burger King and *Beauty and the Beast* incidents when he argued that his surveys showed the VoiceLab ads had an extremely high degree of acceptance. "Out of more than a million impressions we had less than five negative responses," he told the TechCrunch website. "Those negative responses were in the form of critical reviews that mentioned the ads. That's less than a fraction of 1 percent."[17] If Amazon was listening, it was not persuaded. A cynic might suggest that it simply didn't want competition for its own eventual ad network. In any event, on May 21 the company put a second crimp in Marchick's enterprise. This time, it prohibited using simulated voices on the Echo in ways that imitated an "Alexa interaction." This ban struck at the core of what VoiceLabs intended to do, and Marchick threw in the towel. "The May 21 policy change by Amazon really drove home that the market is not ready," VoiceLabs noted in a press release announcing that it was suspending its advertising program. "VoiceLabs and our partners were most excited to introduce interactive advertising that converses with consumers to create innovative experiences. This ability to react to user preferences opens the door to a whole new field of audio advertising, and the May 21 policy prevents this."[18] Marchick shifted his full-time business to helping more than a thousand developers of Google Assistant and Alexa apps to analyze their visitors. But he didn't seem satisfied with that. Half a year later he changed VoiceLabs' name to Alpine.ai. This new business, he told VentureBeat, would help brands and retailers create voice apps that would answer users' product questions and encourage app-based purchases.

Marchick had finally done what Amazon and Google quietly wanted him to do: he gave up pushing the spread of advertising on their smart speakers and instead began encouraging voice-based apps that would lead speaker owners to feel good about their devices. An audit by Voicebot.ai, a website specializing in digital voice news, noted that the number of voice-based Amazon skills and Google actions available to U.S. users more than doubled in 2018. The number of Amazon's voice apps in the United States rose from 25,784 at the beginning of the year to 56,750 at the start of 2019.[19] Google's raw numbers were far lower, but Voicebot.ai found that its 1,719 apps in January 2018 had increased to 4,253 a year later.[20] To help users sort through the rising numbers of voice apps, Alexa and Google Assistant provided app recommendations in response to users' questions, while Amazon and Google created websites to help people find voice apps on specific topics. To encourage the creation of yet more apps, the voice titans reached out to marketers. Although what marketers really wanted was to advertise directly in response to people's questions to Alexa and Google Assistant—for example, "Walmart is .7 miles away and its fresh fruit aisle has a lot of your favorites"—Amazon and Google offered app creation as an alternative, contending that building apps can enhance brand personality. To make that prospect even more enticing, they offered a particularly desirable carrot: the opportunity for a company to harvest data about its voice-app visitors.

David Isbitski is Amazon's chief cheerleader to marketers around these topics. With the formal title of Alexa evangelist, he travels the country encouraging executives to create voice skills, which he claims will allow their brands to reach out to potential customers in ways that traditional mobile phone apps cannot match. An Echo skill, he says, can be thought of as a "contract for conversation."[21] Let's say you want to make a skill about hiking. You come up with two hundred questions a visitor is likely to ask, such as what kind of shoes should I wear, and what makes a good backpack. You give those questions and their answers to Amazon as "sample utterances," and they become part of functions that are called when a visitor speaks. If visitor Mike asks about boots, that is the "boots" function; if it's about the best hiking trails, that is the "best trails" function. Anticipating the range of utterances that visitors are likely to use, Isbitski points out, is often the hardest aspect of working on a skill. Amazon offers a dialogue management program that gives skill creators free help with routine question-and-answer setups.

Another key to developing a successful skill for an Amazon device, Isbitski says, is a choice of voices. He tells me, "I constantly tell brands, not only do you not have to use Alexa's voice, but it might [even] be better for their skill to choose a very different one." A medical site, he suggests, might want to use an older-sounding voice to reflect maturity. Different pitches indicating different genders create different sorts of invitations to engage in conversation with the Echo. "That's not something that happens in mobile," he continues. "But it is absolutely something that happens in a [skills] conversation, because we want to create a connection human to human." Amazon therefore offers skills creators free access to a service called Polly, "which allows you to generate very natural life-like sounding voices with a wide range of speech sounds." Polly offers more than fifty voices to American developers, and along with it they can use, also free, Polly's speech-synthesis markup language. That allows a developer to adjust the chosen voice to match the meaning of an utterance—raising it a bit at the end of a question, for example, or making it whisper when appropriate. Isbitski emphasizes that developers can also choose their own voice for their skills. In 2020, for example, the *Jeopardy!* TV program's skill used the voice of the late Alex Trebek to introduce the game and its categories. The voice of Alexa then spoke the game's questions in the form of answers. Amazon provides the app creator with a certain amount of free space on the Amazon cloud from which it can interact with visitors in the skill. Or a creator can use his or her own cloud, as long as it uses a secure protocol.

Visitors to a skill can purchase things during their visit by using Amazon Pay for Alexa. This allows Echo users to apply the same Amazon account information they use when they buy from Amazon's website. Visitors don't have to log in every time they make a voice purchase on a skill, though users do need to approve the use of Amazon Pay for every voice app they use.[22]

Amazon also allows individuals to set up a secure PIN to prevent unauthorized use. Fairly seamlessly, then, companies can sell through their skills many forms of material items, as well as digital products like premium subscriptions that offer audio programs without ads. Amazon keeps 30 percent of the purchase price for these activities, though charities that set up shop on its smart speakers are allowed to keep the full amount.[23]

Google's approach is similar. Both it and Amazon allow app owners to track their visitors and use the data elsewhere. Marketers know that Google and Amazon can glean a lot more information about users than their individual apps can, because they can monitor users' visits across apps as well as the questions they ask and the items they buy from the smart speaker outside the apps. Nevertheless, the two firms do permit voice-app owners to collect the same insights into a visitor that a mobile app can gather—which is a lot. As long as the owners disclose the activities in their privacy policies (which visitors would likely not notice, let alone read), Amazon and Google allow them to ask visitors to reveal their identities by logging in. Echo and Google Home also offer dashboards to give voice-app owners insights into visitors' activities—what they said, whether their utterances succeeded in answering their needs, the length of time they stayed, and in at least some cases transcripts of what the visitors said. Many developers build even more knowledge about their visitors by hiring firms like Adobe and Dashbot to create tracking tags. These allow the marketer to recognize return visitors even if they haven't logged in, trace voice-app visitors to other platforms, and sell the information collected about individuals to third-party data firms. Overall, it's an odd message to voice-app developers: Amazon and Google welcome the hidden tracking of visitors via apps even as they often won't allow obvious advertising on the same apps.[24]

The strategy of encouraging voice apps from marketers (not just from random individuals) has clearly worked. In 2019, a Forrester consultancy report said that half of the firms surveyed already used, were piloting, or planned to set up "a voice-based digital experience."[25] These creations tended to be aimed at giving knowledge and good feelings to target audiences, rather than direct commerce. Forrester suggested that the increasing number of voice apps, along with the rising number of U.S. households with a smart speaker—28 percent at the start of 2019, projected to rise to 50 percent by 2022—was making firms believe that they had to participate. Yet the authors of Forrester's report were unenthusiastic about voice apps' value, writing that "most voice interactions today are mediocre at best" and perceived by customers as "unfamiliar, unreliable, or unhelpful."[26] It is possible to get voice apps right, but not many firms were succeeding.

Joe Maceda spoke for many ad agency practitioners who insisted that Amazon and Google's attempt to satisfy marketers with voice apps instead of ads on smart speakers wouldn't work.[27] Maceda is a senior executive at the huge Mindshare media buying agency. "Skills are essentially the new mobile apps," he told me, noting that others have said the same thing. "Ten years ago we all rushed to create thousands of mobile apps for brands and almost all of them now sit somewhere buried in the App Store and haven't been updated in a decade because they weren't doing anything the consumer needed." Most consumers, he said, can use perhaps "ten mobile apps or something like that." With a device such as a smart speaker, "when there is no visual interface to toggle through your skills," he believes the number is even smaller.[28]

The executives I interviewed on this topic who weren't as pessimistic as Maceda were nevertheless cautious. Michael Dobbs, a senior vice president at the multinational digital advertising agency 360i, said he and his colleagues believe that a voice app, potentially one with a screen such as Amazon's Echo Show, "is a better way to communicate with the consumer than with a website that is a purely screen-driven experience." Yet he noted that "a lot of the [voice] experiences that have been designed today are pretty poor" and offer little of genuine utility to customers. He also said, as others did, that Americans' usage of skills and actions, apart from some games, "has been rather low so far."[29] Kirk Drummond, of the Drumroll ad agency in Austin, Texas, saw the technology of voice apps as a work in progress that might eventually have real value for his clients. Drumroll works only on Amazon skills (because of the "dominance of its installed base") for firms such as Thunderbird food bars. Though users can purchase bars directly from the skill, Drummond said that "we don't look at these skills predominantly for commerce" but for "reinforcing brand . . . perceptions" and turning the brand "into something that some-body has an emotional connection to . . . and would actually go advocate for their success." Even though he works with clients to explore the voice-app phenomenon, he's "not sure what it means yet in terms of value to my [client's] brand."[30]

Pushing back against such wariness, Amazon and Google have insisted that skills and apps will eventually become valuable to marketers for selling and for audience data. In 2017, Google's head of advertising and commerce, Sridhar Ramaswamy, suggested that Google itself could gain a lot from the voice apps by taking a cut of each purchase that customers asked their Home device to make from Costco, Whole Foods, and other partner stores. But by 2018 it wasn't at all clear whether the commerce approach was working. In early 2018, OC&C Strategy Consultants estimated, based on a survey of smart speaker owners, that Amazon and Google had sold through U.S. devices about $2 billion worth of products, mostly "stand-alone lower value items," and predicted that voice shopping would soar to $40 billion by 2022.[31] But other public surveys in 2017 diverged widely on the number of Americans who had even ordered anything via a smart speaker, from a high of 57 percent to a low of 19 percent. The tech site The Information found evidence that sales were far lower: "according to two people briefed on Amazon internal figures," it reported, only about 2 percent of Echo device owners had purchased anything through their smart speakers that year.[32] At the end of 2018, eMarketer concluded that "few people regularly make purchases through smart speakers," and expressed the conventional view that "for the time being, voice commerce is probably best-suited to replenish goods, where the shopper already knows what they want and can utter a simple command to reorder that product."[33] Half a year later, Greg Sterling of the Marketing Land trade site reported that Google was "signaling high hopes for the Assistant to become a booking and commerce platform." But, he cautioned, "users have yet to respond to these capabilities at any level of scale."[34]

Amazon and Google didn't comment on the figures publicly, or present their own. Their silence invited several interpretations, among them the conclusion by a journalist reporting on Ramaswamy's 2017 talk that "making money is not the priority [for Google Home] now." That also explained why the companies were restricting advertising. The aim, the journalist suggested, was to cultivate users and market share before opening the system up to profits.[35] Amazon's David Isbitski went so far as to tell me that Amazon had no profit motive for its smart speakers. The company wasn't making money from selling the devices, he said, because they're so inexpensive. It also doesn't currently make much money from direct product requests ("Alexa, order batteries") or from skills purchases that use Amazon Pay to complete a sale. (The skills can also use other payment systems.) And this, he insisted, was not going to change. Amazon would not profit in the future from Echo's services or by gathering in loads of revenue from advertising. "It's not about making money," he told me. "It's about making customers' interactions with technology easier. . . . Think of Alexa as a startup funded by this VC called Amazon.com." Yet startups are open about their plans to make money; that's why venture capitalists fund them. Pressed about Amazon's long-term business goal for the Echo, he said, with a bit of frustration in his voice, "It's hard to imagine if you're not in a culture that is so customer centric that even just saving a fraction of a second for a customer is worth—, is priceless." When told that sounds like a public service, he answered, "There's more things than just money. I'm not in any meetings that say we're doing Alexa to make money."[36]

Google's Ramaswamy did project an endpoint to what he seemed to acknowledge was presently a loss leader. "We are very focused on getting customer experience right first," he said in 2017.[37] While not closing the door to ads, he depicted the ad-free decision as a move to prioritize marketers' use of voice apps. "'More transactional than ads' is how I would like to think about [Google Assistant] right now," he said.[38] He repeated his company's position that the *Beauty and the Beast* insertion was an experiment, not a glimpse of where the company wants to place its future emphasis.

The difference between Isbitski and Ramaswamy's positions showed up more broadly in an interview with the *Wall Street Journal* in 2019. When the reporter asked Google and Amazon about the future of their bans on advertising, the Google spokesperson paraphrased the marketing chief: "Our focus right now is on creating a great user experience and making sure that the Google assistant can help you get more things done in your day." Amazon, the reporter wrote, "declined to comment."[39]

Many marketers, annoyed at their advertising being relegated to voice apps that are needles in huge smart-speaker haystacks, take such statements by Google, and even a non-comment like Amazon's, as hints that they will eventually get what they want: the ability to purchase ads in response to what people say to Alexa or Google Assistant. Some executives believe the shift to voice-targeted ads is an inevitable part of an ongoing transition to voice across all digital devices. Lubomira Rochet, chief digital officer of cosmetic giant L'Oréal, phrased it this way: "The question is: Will voice surpass fingers in the way we type into the interfaces? Will people choose to stop typing SMS [short text messages] and speak them? The day that happens, marketing and commerce will become voice-led."[40] *Advertising Age*'s senior editor for media and technology, Jeanine Poggi, sees this beginning to happen. Among her predictions at the start of 2019 was that "brands will invest in voice marketing strategies in a big way." They "can't afford not to," she argued, given that more than half of U.S. households are expected to have a smart speaker by 2022.[41]

A merchant's ability to reach individuals based on what they say and how they say it—linked to other data about who they are and what they are like—fits easily into contemporary frameworks for buying and selling ads. Over the past couple of decades, digital publishers, advertisers, their agencies, and tech firms have developed marketplaces where publishers ask for bids from companies that want to reach individuals with highly specific characteristics who have come to the publishers' websites, apps, or other locations. Many publishers, Google and Amazon included, choose winning bids from these companies based not just on the cash promised, but also on their predictions about whether the people targeted by the ad will act on the sponsor's message. The bidding process increasingly takes place in fractions of a second, at the very moment the targeted individuals are going into the website, app, or smart TV—and the same could happen with the smart speaker. After the commercial message is played for the user, the publishers and the advertisers use various metrics (including product purchases, clicks on links, and phone calls) to measure success, and the advertisers sometimes have the ability to continue reaching out to those audience members. Amazon and Google are market makers in this approach to digital advertising. And it's easy to see how this same bidding system for audiences—currently based on what users are typing, viewing, or listening to—could be adapted to include requests based on what these users are saying and how they are saying it.

Ad executives I interviewed and those quoted in the trade press were frustrated that such a marketplace doesn't now exist with the Echo and Google Home, especially in view of the huge storehouses of data, including voiceprints, that Google and Amazon are building. But they are determined to be ready if and when the situation changes, and they are starting by trying to understand how what people ask for translates into non-sponsored (or "organic") responses by Alexa and Google Assistant that mention their products. Marketers believe that in traditional text search, how well a brand does in organic results makes a difference in determining whether that brand's paid advertisements are chosen to be shown to individual users. "I think we can assume that eventually brands will have the opportunity to buy their way into voice results," Maceda said. "But, just like in text search today, being the number one organic result puts you in a better position to be a better bidder for paid results. So that's why we are putting all— well, not all but a vast majority—of our focus on building organic performance in voice search."[42]

What advertisers do to help particular websites appear among the top results when people ask a search engine a question is called search engine optimization (SEO).[43] Amazon doesn't appear to disclose ways to help companies move their products up in its store's search rankings. Google and Microsoft, the dominant web search firms, offer guides for advertisers on how to prepare their websites to succeed in organic search, but they keep many particulars of their search algorithms secret.[44] They also continually change the algorithms in response to changing consumer query patterns and various attempts to game the system. The details the firms do disclose are complex, and the stakes for marketers are so high that in the past two decades a mini-industry of consultants has emerged to help advertisers shape their sites to encourage good SEO results. The increasing use of voice on mobile search engines and smart speakers has led some in the search business to believe that Google and Amazon are changing the search rules because, as one data scientist put it, "people type and speak very differently."[45] Building a web page to optimize for how people ask questions by voice could bring a huge competitive advantage if a company's rivals are still optimized only for typed queries.

Marketers and their consultants, then, are trying to understand these talk-type distinctions to improve their rankings in voice-search results offered by websites, phone apps, and smart speakers. "People are trying to understand, 'OK, I've spent ten years getting my SEO strategy together, making sure I am relevant in searches," said Pete Erickson, founder of Modev, which specializes in conferences for software developers. "'And I do all these things. But now that voice-first is here, what happens to my search results? Am I relevant? What's happening?'"[46] Search consultant Bradley Shaw wrote dramatically in 2019 that many SEO practitioners feel "shell shocked" by the changes and are "scrambling to adapt."[47] The web is filled with articles about how to populate websites with features that voice-search engines look for.[48] One example: Recognize that people talk longer and more casually than they type, so the words that sites place on web pages—which search engines monitor to determine relevance—should fit a more colloquial style. ("When does Walmart open?" rather than "opening times: Walmart.") Another example: use terms such as large, medium, and small to describe a jar of, say, mustard, mayonnaise, or ketchup—because this is the way most people speak—in addition to the required listing of fluid ounces.

To large media agencies, these widely shared ideas are table stakes. They have for several years been mobilizing substantial resources to help their clients decipher obscure aspects of organic search so that they can win auctions for paid voice-search ads whenever Google and Amazon finally roll them out.[49] For example, 360i, a subsidiary of the giant Publicis agency holding company, formed a specialized Amazon marketing practice led by Will Margaritis, who had previously worked in e-commerce at Amazon and L'Oréal. *Advertising Age* reported that 360i has developed a "proprietary Voice Search Monitor" that "reverse engineers algorithms used by Amazon, Google Home, and Apple's Siri to help marketers better understand them."[50]

Margaritis's Atlanta headquarters is filled with soundproof boxes and microphones that ask hundreds of questions of each assistant and record the results. The idea is to discover how smart "the devices" are in answering, where the gaps are in their knowledge, how that's changing, and how clients can use the answers to affect their voice search results. A lesson his colleague Michael Dobbs highlighted from the research zeroes in on what marketers call the "implicit discovery" of voice apps. Implicit discovery happens when a person is led to a voice app not through direct search but indirectly, by asking the assistant a question and the assistant then pointing them to the app.[51] For example, Dobbs explained to me, his firm tried to figure out how to get Google to suggest the user go to Starbucks's voice app after the user asks the Google Home for a coffee shop or café with free WiFi—a feature not unique to Starbucks. The larger question is: what information does a store's voice app need to give a visitor so that Google will privilege that store's app over others in its organic voice results? Dobbs says the team learned that Google uses several elements of an app in deciding which retailer to highlight, but the one that stands out for "coffee shop or café with free WiFi" relates to the app asking if the visitor wants to use GPS to find the retailer's nearest location. If the app includes that choice (or "utterance") in its toolkit, Google's algorithm is likely to point to the app. Findings like this, Dobbs said, give advertisers power with respect to Amazon and Google's voice assistants. "We feel like building out your [voice apps'] training phrases and sample utterances in a smart way together with some machine learning on our part to facilitate that, we can increase the odds that you will get your impression [that is, mention of the voice app] for Starbucks or some other brand."

At an even further edge of personalization, ad practitioners from the biggest agencies are trying to learn how to use data from a customer's voice as a means of persuasion. These executives know that both Google and Amazon's voice assistants are able to recognize unique voices and present different answers based on who is speaking. They also understand Google and Amazon's ability to infer characteristics from unique voices and speech patterns. Yet few in the industry believe that Amazon or Google will give them access to people's actual sounds on smart speakers in the next few years, if ever. An alternative, several noted, is to retrieve voiceprints elsewhere—for example from websites, podcasts, and store kiosks. Mike Dobbs told me that part of "the stuff we do" at 360i is to try to learn how to infer meaning from the voice of a person searching for a product. "How do we evaluate the input of the searcher? What's the tone and the mood—what can AI do with that? Can the AI do this in a more persuasive way?" He didn't have answers, and he allowed that "maybe it's more of a reality than I can conjure right now," but the work is proceeding.[52]

I asked six executives to forecast a timeline for voice profiling. Publicis's Rishad Tobaccowala was dubious that it would ever emerge because he didn't think the public would stand for yet another worry about online privacy. The other five all gave the opposite opinion—that in one form or another, the profiling of voices as a way to understand and target the customer is inevitable.[53] A few executives brought up the public's nervousness about data capture and changing government rules about privacy as factors to consider when introducing voice profiling, but they added that those obstacles could be overcome as people began to see the benefits. Will Margaritis, Dobbs's colleague at 360i, said that, due to privacy concerns, he doesn't think society is ready for targeted advertising based on voice profiles, but it will be, "probably soon."[54] Austin Arensberg of Scrum Ventures, an early stage investment-capital firm, suggested that, except for contact centers, marketers have not done much with voice profiling because it is too new and too expensive, and for the moment, not likely to work well. He predicted this would change sometime in the early to mid-2020s.[55] Kirk Drummond said that "without a doubt," voice assistants would eventually sense whether a person is upset or angry and change the tone of a targeted ad based on that data, which he called tonal intelligence.[56]

Mindshare's Joe Maceda stressed that customers should be given a choice in this activity, but he was sure most would agree.[57] "I would assume," he told me, "that even without regulation there would be some level of consumer opt-in to that level of depth of understanding that you are providing your assistant to start giving you recommendations based on entirely nonverbal [cues]." Getting that approval, he proposed, wouldn't be difficult. He expected that "consumers are going to increasingly be willing to give up their privacy." Times have changed, he stressed. "If you walked into Blockbuster video in 1992, and the guy behind the counter said, 'Hey, I know exactly what you've been watching on TV for the last two weeks—here are some movies I can recommend—you'd be freaked out. But that's exactly what you get from Netflix, and nobody seems to blink an eye. Once you take the humans out of the equation, I think people are comfortable giving up some privacy when there is a greater benefit for personalization."[58]

Accepting this proposition, people at Mindshare's NeuroLab have been working with the company's Global Strategies subsidiary to figure out ways to apply emotions from people's voices to understanding them as they use personal assistants. NeuroLab, launched in 1999, says it "uses medical-grade EEG (electroencephalogram) and GSR (galvanic skin responses) to measure second-by-second, non-conscious neurological responses to brand stories and media." The NeuroLab also gives subjects word-association tasks in an effort to understand their biases regarding certain brands and images, and it conducts quantitative surveys with those aims in mind. Mindshare executive Janet Levine told me that NeuroLab's focus on emotions comes from a desire to add value to current data that their clients use.

We realized that the way that we advertise—and by *we* I mean the industry—we are reaching people through external factors. You [the advertiser] know that I [the consumer] am sitting in this location in New York, I bought this type of clothing online, so I must be served by this type of ad. But all of this, these cuts that we do—demographic, behavioral, location shopper, even competitive targeting—a lot of that is external. Which sort of begs the question: what is the next thing that we could be going after? And that's emotion.[59]

Early NeuroLab studies have focused on what its researchers say are the intense emotions that stories about brands evoke in people when the stories are conveyed through sound rather than sight. NeuroLab leaders don't yet have the ability to detect the feelings in a request someone makes to a voice assistant, but they are headed that way. Levine said, "I specifically asked that question the other day. It's not quite there yet." But "what they were saying to me at a high level was that is the future. That is absolutely where we will go. But right now . . . we are in the early stages of that as an industry. The first piece is understanding how people interact with their voice assistant. Later on, it will be, 'Were they asking the question because they were happy? Do they have a little tension in their voice, and it was a long day and they're tired or maybe getting sick?' But that's more in the future."[60]

Many in the contact center business would point out that they are doing this already. Arafel Buzan, NeuroLab's co-lead, argued that one of the things that the firms creating the call software get wrong is that they have a simplistic understanding of the range of emotions.[61] Those software creators would undoubtedly reply that their emotion-tagging of voices succeeds for clients. They can show numerical evidence that intelligent call routing driven by personality or emotions substantially increases caller satisfaction and upselling. Ironically, many advertising executives who know little, if anything, about the personalized phone world represent companies that separately hire experts to carry out 800-number contact-and-analysis activities. An ad executive asked about this stated simply that the arrangements were made by different parts of the company, and no one had made the connection.

The contact center analytics company Mattersight did see a link, as well as the potential profit in it. In 2017 it received a U.S. patent for a method by which "a digital assistant, such as Siri, Alexa or Google Home, . . . will tailor its interactions with users based on their personalities."[62] VentureBeat reported that "using artificial intelligence to recognize your personality within 30 seconds, Mattersight wants to deliver personalized ads and bots to Amazon Echo, Google Home, and other voice-controlled devices."[63] The article's author, Khari Johnson, explained that "Mattersight assigns personality profiles to the people who call major brands, and it wants to bring that same service to intelligent assistants, using voice biometrics as an identifier instead of phone numbers." Johnson quoted Mattersight executive Andy Traba as saying voice assistants are simply a form of bot that can be controlled in the same way that call centers control their own AI bots.

Traba told me he and his Mattersight colleagues are confident that companies will eventually use voice identification and other speech bio-profiling for ad personalization.[64] Yet most marketers, he argued, are not prepared to deal with speech data, because such data are far more complex than the demographic and behavioral inputs they typically use now. He expressed surprise that figuring out how to understand speech is still "not at the forefront" of firms' strategies given "the amount of money and capital that is within the system [and] the increase of voice and voice taped interactions." But he thinks this will change. Echoing Mindshare's Joe Maceda and Janet Levine, he argued that advertisers will chase voice profiling in earnest when executives can no longer get enough new competitive insights from their traditional ways of thinking about their customers. Marketers' ability to extract value from the older forms of data will inevitably "start to plateau," and at that point, practitioners will "obviously have to start looking for more ways to create value." Executives, recognizing the utility of voice, will find ways to target individuals based on their speech, and "the contact center and personal assistant businesses will converge."[65]

Marketers' optimism about improving their ability to exploit voice for personalized messages has been dampened by concerns that the voice titans will make it difficult for marketers to apply their knowledge, or will be unreliable dealmakers with future voice-driven marketing initiatives involving Alexa or Google Assistant. Some even worry that Amazon or Google will appropriate the marketers' audience data, including voiceprints, for their own use. Such concerns have led some marketers to create their own intelligent agents for exploiting customers' speech.

The overriding dilemma that concerns advertising executives is what they call the "voice-first problem." They fear that with speaker-based products, which privilege voice rather than text commands and queries, the number of ads will drop sharply—even on screens such as the Amazon Show or Google Nest Hub. Joe Maceda of Mindshare calls this circumstance the new "voice shelf." The range of products in any one category, he adds, "is probably going to be two or three brands. If you are the third or the fourth brand in a category, there is significant risk."[66] Many smart speakers have no displays at all, of course, and executives worry that even if voice marketplaces became available, Google or Amazon might not allow more than one voice ad in answer to a user's question (though sometimes they may actually offer two or three choices). The possibility that only one spot will reach a potential customer—a situation called position zero—has become a fixation among advertisers. They fret that the voice assistants may offer users only a single choice, and further, that the choice will reflect what Amazon and Google data suggest the person is already buying. Maceda calls this new circumstance "incidental loyalty." People may continue using a brand, he says, "not because that consumer loves the brand's position or even prefers its product performance." Rather, "it's loyalty to a brand because they have taken themselves out of the decision-making process and handed it over to a third party—in this case some sort of digital assistant or voice assistant."[67]

Such activities may often privilege brands with which Amazon and Google have monetary ties, says Mykolas Rambus of the Equifax DDM data consulting firm. This already happens, he argues, in organic voice search. "Take AmazonBasics, its store brand, as an example. When someone asks 'Alexa, please order batteries,' Alexa doesn't say 'would you like me to explain the options,' right? Alexa usually gives your shopping list . . .

AmazonBasics, like AmazonBasics batteries." Often, he acknowledges, Amazon and Google suggest products based on a customer's previous purchases—what some might consider organic results that don't reflect the search firms' commercial bias. But he said that scenario is also a problem. "It's certainly good enough if you are the incumbent brand. But if you're trying to conquest a customer [get him or her to switch brands], that's not going to be sufficient. That sort of suggests that the algorithm is biased towards the things people buy even if there is a different product that might be a better price or greater value that would be more apt to the consumer's preferences."[68]

In an environment where the voice titans have no interest in helping ad practitioners win customers in voice, marketers fear that the best they can do is to create content on their voice apps, phone apps, and websites that may lead Amazon and Google's search engines to offer their product as an organic answer to a user's search query. Google and Amazon "are not helpful at all," said Forrester consultant Brandon Purcell, in part because "they don't want to completely expose how these algorithms work because that's a big part of their IP [intellectual property]."[69] Michael Dobbs of 360i agreed that Google and Amazon are husbanding the keys for themselves, creating an "impasse." He complained that "we can't even get the voice-query data [about what smart speaker owners ask, and how that leads to particular brand-related organic answers]."[70]

Some marketers suspect that Google and Amazon may be harming them in other ways, too. In particular, they worry that Google and Amazon will find holes in contractual agreements that will allow them to use the firms' customer data. Recall from Chapter

3 that an auto executive told me "somebody at Google" acknowledged that "if carmakers really knew what's behind Google, it would really scare them. They don't even know what Google is doing."[71] The notion that Google—and Amazon—can't be trusted with a partner's data also came up in a discussion involving the Mars Agency and client Estée Lauder. Fresh from its success in using Alexa for Business to help sell whiskey in Manhattan's Bottlerocket liquor store (see Chapter 3), Mars lined up the cosmetics firm for a similar setup in a department store, where a smart speaker would help shoppers find Estée Lauder cosmetics appropriate for them. The difference this time, according to Mars executive Ethan Goodman, was that his client nixed the use of the Echo and required Mars to use Google Home instead. The Lauder executives were worried that Amazon would capture the voice interactions moving through Alexa to learn about Estée Lauder's customers and their concerns, and use that information to inform its own cosmetics sales strategies. "We're hearing that a lot more frequently from retailers," Goodman added. "They don't want to have anything Amazon in their store." He pointed out that "Amazon says that [with Alexa for Business] they provide us with a portal that allows us to log the data, and they're telling us they don't access that"—but, he added, companies don't believe it. Goodman himself said, "I'm sure the truth is somewhere in between."[72]

This corrosive attitude is even leaking into mainstream media. "The data appetites of Facebook and Amazon are a serious impediment to companies considering these platforms [Facebook Portal and Amazon Echo] for customer service," wrote two reporters for MarketWatch. "Why would a retailer set up shop on Alexa when Amazon, likely the retailer's biggest competitor, might be eavesdropping on customer conversations?" The article quoted the chief technology officer of Mobiquity, which built the Alexa skill for Butterball: "Everyone in retail is afraid of Amazon. It is an instantaneous fright about the intelligence gathered on their business from using that voice channel. I have some customers really, really scared about that, and it is stopping them from doing development. They are at least a year and half behind what their original plans were because of their paranoia that intelligence will be gathered about their business and used against them."[73]

It's not surprising, then, that while tens of millions of people reveal themselves to smart speakers daily, many marketers are wary of using them to engage with customers through voice apps. Some companies' response is to "skip the smart speaker" and create their own AI-powered voice chatbots on the web or for the phone, often using tools from Microsoft.[74] Bret Kinsella, founder of a website and newsletter about the voice industry, believes this is a key trend to watch. He wrote that "companies want control over the user experience and session data. . . . If forays into Alexa and Google Assistant convince companies that voice is an important consumer engagement method for the future, they must quickly confront the question of how much control they are willing to cede to a third party and [to] which third parties."[75] Some companies are beginning to create firm-specific voice assistants on websites, as well as tablet and phone apps, that are totally separate from the big voice firms. Kinsella pointed out that nowadays, creating a compelling assistant doesn't mean initiating speech recognition and natural language models on levels that would compete with companies as advanced as Google, IBM, Amazon, Nuance, or NICE; that would take a billion dollars or more. A few companies may take that route, but most will find that they can instead use a toolset created by Nuance, SoundHound, or another "white label" company to build their own assistant—one that they can brand and, even more important, one from which they can acquire and keep user data.[76]

The ad-supported audio streaming service Pandora, for example, uses an assistant called SoundHound. Pandora's goal in implementing SoundHound, beyond encouraging users to speak commands for particular types of music, was to allow them to respond verbally to ads. Because Pandora listeners don't normally look at the screen, they aren't likely to click on ads or do other things that advertisers expect for visual persuasive messages. SoundHound, Instreamatic's interactive ad technology, had a solution: it asked the user questions (for example, "Hey, do you want to know who makes the best pizza in town?"), and when listeners answered, Pandora had the proof it needed to satisfy advertisers that they had engaged with the ad. Eric Picard, Pandora's head of product management, sees this assistant's new capabilities as transformative: "Voice interactivity has already changed the way consumers interact with brands on smart speakers, and we believe voice will change the very nature of the way consumers interact with brands on Pandora."[77] The company began live testing of the interactive voice ads in December 2019. A group of large advertisers took part, including Ashley HomeStores, Comcast, Doritos, Nestlé, Turner Broadcasting, Unilever, and Wendy's.[78]

Another company deeply involved with a standalone voice helper is Bank of America. "Meet Erica, your virtual financial assistant," heralded the bank in late 2019, on a promo page for its new helper, the first voice-interactive financial chatbot.

Created by the bank itself, it was introduced to the industry in 2016 and rolled out to Bank of America's approximately 45 million customers in 2018.[79] Users can tap buttons to communicate with the assistant, or they can type or speak. "Erica can help you pay your bills anytime, from almost anywhere," according to a video on the promo page. It shows a young woman seamlessly switching from typing on her phone to reach Erica and asking it to do bank-related things. For example, she types to ask Erica to lock her card after thinking it is lost, and she transfers ten dollars to a friend by instructing Erica via the phone keyboard. Then, when she finds her credit card in her jacket, she uses her voice to say, "Erica, please unlock my credit card." A female voice responds: "No problem. Your card is now unlocked and ready to use."[80]

The company noted that the female name is a shortened version of "America" in the company's name, though a bank representative conceded in 2018 that Eric might have been just as appropriate. The rep noted only a few complaints about the assistant's gender and suggested that in the future people might have a choice between the male and female versions.[81] By late 2020 it hadn't happened. The video, avoiding any discussion of that issue, emphasizes that unlocking cards and making payments are only some of an enormous number of tasks that Erica can accomplish for Bank of America customers. The bank's website emphasizes that Erica can help clients find transactions (for example, "Can I see transaction activity for cable satellite?"), access account information ("When is my credit card payment due?"), pay bills ("Pay a bill"), send and receive money ("Show my scheduled transfers"), organize spending and budgeting ("How much money did I spend on Saturday?"), secure credit and debit cards ("Report my card lost or stolen"), view rewards, benefits, and FICO scores ("Why did my credit score change since last month?"), and contact the bank ("Schedule an appointment"). A 2019 press release announced that Erica had extended its reach so that it can now also give "personalized, proactive, and predictive" advice by using "the latest in artificial intelligence (AI), predictive analytics, and natural language processing." Among other capabilities, it can notify clients when their spending pattern could take them to a negative balance in the coming week.[82]

Much as with Alexa, Siri, and Google Assistant, not all users were pleased by the voice app. One person sniped that Erica couldn't understand her request after several tries. Bank technology personnel acknowledged the difficulties of having the app respond to the wide range of slang around money issues. (An early realization was how many people use the word *dough* instead of *money* when they speak rather than type on their phones.) Nevertheless, the bank attributed the voice app's rapid growth to its success with personalization. In May 2019, a year after its rollout, Erica had accumulated 7 million users and completed more than 50 million requests. Half a million new users were coming on board each month.

Erica is a sophisticated version of the dedicated voice apps that Bret Kinsella is sure will proliferate. During 2019 the idea seemed to be spreading. Other organizations announcing plans to create a custom voice assistant included the British Broadcasting Corporation and the retailer Ikea. In suggesting that these developments portend a trend, Kinsella emphasized that he doesn't see developers of such assistants as trying to provide all the functions of Alexa or Google Assistant. Many will instead reflect their brands' identities, focusing on specialized skills. Voice-conference organizer Pete Erickson agreed, writing, "I believe the price-point for developing and releasing custom, (almost) single use devices is down to a level that will make [Kinsella's] predictions even more prevalent."[83]

Future-leaning ad executives suspect that despite their current tensions with marketers, Amazon and Google will reach a point where competition with one another, as well as with Apple, Facebook, Samsung, and other media firms, will push them toward encouraging advertising based on people's speech patterns and vocal sounds. To Pete Erickson, that scenario is ultimately in Amazon and Google's interests.[84] "I think they realize," he told me, "that if they don't get this right with regard to the market they are going to be stranded. Because people get tired if they're looking for things and are only hearing what Google wants you to hear, for example, or they are only hearing about Amazon products and not third party products." As Forrester consultant Brandon Purcell notes, "All marketers I talk to are interested in the omnichannel customer journey."[85] They hope they can piggyback on voice-first devices to follow where their customers go in their homes, their vehicles, stores, and even hotels. And they expect that eventually Amazon and Google will share voice data with them that will allow them to bring personalization to a new level.

Yet the tensions around Amazon, Google, and their smart devices have opened up a perhaps unexpected new domain in the spiral of personalization. Though there is no evidence that

Bank of America or Pandora is now using voice profiling, they are at the leading edge of developments that promise to move tailored targeting and bio-profiling into a higher gear. While the multiplication of voice apps might prove useful to individual firms, the competition it brings may hasten discriminatory practices based on individuals' speech and sounds. Marketers suggested to me that Amazon and Google are moving slowly with their voice-intelligence advertising activities because they share a large public spotlight, and many of their practices affect people's sense of privacy. But much as in the contact center business, smaller and lesser-known firms might get away with more aggressively judging people by their voices. A recurring prediction is that after 2020 more than half of all searches in all devices will be through the human voice.[86] As the spiral of personalization goes into overdrive, consultants may raid the scientific literature to sell the ability to extract from voice a broad gamut of inferences about people's emotions, person-alities, gender, social status (read income), ethnicity, weight, height, and even illnesses.

A profoundly important new industry is preparing to sweep across society, and there is little public discussion about the future we might want for it or the social policies that would get us there. This conversation becomes especially urgent when we consider the long-term harms that could result if voice profiling and surveillance technologies, put in place by the voice intelligence industry, are used not only for commercial marketing purposes, but also by political marketers and governments.

6 VOICE PROFILING AND FREEDOM

Voice profiling is a gateway drug to a new era of hyper-personalized targeting. Since Apple introduced Siri for the phone in 2010 and Amazon debuted Alexa in 2014—with Google Assistant catching up and Chinese players moving forward since then—speaking to a device and expecting a relevant response has become an ordinary experience for billions of people. Few of us realize that we are turning over biometric data to companies when we give voice commands to our smartphones and smart speakers, or when we call contact center representatives, but that's exactly what is happening. In each of these situations, our voice impulses are converted into binary code, and although nothing in civilization or the laws of economics requires that those zeroes and ones be exploited, a robust and growing industry has developed to do just that. From the home and the car to the store, the hotel, and beyond, companies devoted to personalized marketing are gearing up to add what individuals' bodies say about them to more traditional demographic, psychographic, and behavioral tags.

Consider what would happen if only a bit more of what is taking place in contact centers began happening in the home, and in cars, stores, hotels, and schools. Our worlds would increasingly be filled by offers—not necessarily explicit ads—based on our putative emotions and sentiments. Need help while in the car? If so, how do you know that the intelligent agent is not responding to you based on your voice profiles? Will companies want to deal with you when you sound irascible? Will they answer your questions quickly or linger to offer you discounts? Will the sound of your voice open or close the door to certain deals? Will firms link physical characteristics that they inferred from your voice last week to what you buy today—and combine that data with other information they have collected—to draw conclusions about your health, and so about the benefits and risks in striking up a long-term relationship with you?

The possibilities are endless, and these examples are likely just the tip of a huge discriminatory iceberg. We're already subject to differential offers and opportunities based on various facts about us—such as our income, where we live, our race and sex, and other attributes. Voice profiling adds an especially insidious means of labeling us. We could be denied loans, refused insurance or have to pay much more for it, or turned away from jobs, all on the basis of physiological characteristics and linguistic patterns that we typically don't change and whose existence is certified by a science that may not actually be good at predicting behavior. What if voice profiling tells a prospective employer that you're a bad risk for a job that you covet—or desperately need? What if it tells a bank that you're a bad risk for a loan? What if a restaurant decides it won't take your reservation because you sound low-class (read Black or Hispanic, though the algorithm supposedly corrected for that), or too demanding, or somehow not cool enough for its image? What if a public advocacy organization won't take your donation because its algorithms profile you as gay? Discrimination through voice profiling can be extremely subtle and hard to detect—and thus hard to fight. The problem may be compounded when digital thieves enter the picture. They may steal corporate profiles based on voice—and in some cases voiceprints themselves—and use them for malicious purposes that could range from trying to steal your identity, to spreading unfavorable ideas that companies have about you, to extortion.[1]

Even if shoppers are only dimly aware that these activities are widespread across several industries, they may start to worry about their position in the marketplace and to suspect that the system of commerce is stacked against them. They may also begin to worry that opening their mouths anywhere in public may result in unwanted inferences about them, because microphones are everywhere. As Echo and Google Home were first gaining popularity, reviewers suggested that users push the off button when discussing topics they didn't want the voice assistant to know. Turning the device off, however, takes away the spontaneity that is at the heart of the assistants' seductiveness; when on, the assistant is always open to a question or command from across the room. So people leave it on, leaving themselves open to voice surveillance that statements and patents from Google, Amazon, and others indicate will lead to discriminatory treatment in the American public sphere, where the self and shopping are defined together.

The industry is at too early a stage to try such tactics, and companies are currently wary about moving forward quickly with bio-profiling. But they have the technologies and the patents, and some have begun to use speech and voice analysis in ways that point to a future of treating people differently based on their bodies. Moreover, as we saw in Chapter 5, the proliferation of voice assistants may hasten these activities. And if that kind of marketing discrimination isn't problematic enough for those who value choice and autonomy, consider that in the past, strategies developed for consumer marketing have rarely stayed limited to that arena. As we will see, the applications currently being explored for using voice-profiling tools in political campaigns, border policing, and even prison control suggest that voice intelligence's influence on individual and social freedoms could be extensive and far-reaching.

How would such a future materialize? It all starts with users giving permission. Companies will gain customers' "OKs" by promoting voice-first technologies seductively to them, exploiting their habituation to these technologies, and by not quite explaining how personalization will work. Until now, too, explicit authorization hasn't been necessary. The most advanced voice recognition and profiling activities currently take place within the contact-center business, and voice intelligence firms specializing in contact centers have been almost completely ignored by policymakers and the popular press when it comes to surveillance. The contact-center industry, which connects phone callers to companies at a time when they are often highly stressed regarding those firms' products, has been a perfect venue for perfecting strategies for authenticating a voice, deciding what sentiments it reflects, determining emotions based on tone and other voice qualities, and then crafting a response based on those sounds and word choices. The firms have learned how to embrace certain customers, tamp down their anger, or make them go away without causing trouble. As I noted earlier, people in the industry told me that the "permission" they get from callers to carry out this activity takes place when a recorded voice recites an ambiguous sentence like "This call is being recorded for training and quality control." When voice firms treat individuals differently based on their vocal and linguistic qualities, and merge this information with data about their expenditures and other relationships with the firm, no one outside the contact center knows it is happening—not even the customer who made the call. It's no wonder that both Amazon and Google have entered the business: contact centers allow them to refine their profiling software in an arena that doesn't have nearly the public scrutiny of the smart speaker or mobile phone.

Public scrutiny around smart speakers has forced Amazon and Google to present trustworthy images around their data use. But as we have seen in Chapter 2, even in those realms their stances are muddy. Amazon, for example, openly admits that it integrates what it learns from Alexa-assisted purchases into its databases about users. As of late 2020 Amazon spokespeople were still telling the press that the firm doesn't use actual voice data for marketing, but the company's privacy policies make no such assurances and even suggest that it reserves the right to do otherwise. In its privacy notice, Amazon presents the "voice recordings when you speak to Alexa" as "examples of information collected."[2] The inclusion of voice recordings in their entirety as information would suggest that the company reserves the right to analyze the tone and speech particulars of what people say, even if Amazon is not explicit about this. Google, for its part, states in its Nest privacy policy that it keeps individuals' audio recordings related to its home products separate from advertising and doesn't use them for ad personalization. But the privacy policy also states (in an answer to a question about ad personalization that you have to click on to read) that it does use the *transcripts* of what people say to its Assistant for targeting and profiling.[3] Another exception: Google's general privacy policy explicitly states that it collects and uses for personalized targeting "voice and audio information when you use audio features" of its technologies—which would include Google Assistant on the phone, if not on the Home speaker.[4] This leaves a lot of room to analyze people's voiceprints, as well as transcripts of what they say, for inclusion into the larger profiles that the company keeps on its customers.

Then there's the ambiguity surrounding the voice apps (or "skills" and "actions") that Amazon and Google tout as reasons to use their smart speakers. Several people I interviewed have affirmed that when firms create voice apps, they receive transcripts of what people say to those apps through Alexa or Google Assistant. Firms with voice apps also have the right to collect a raft of other information about their visitors, as long as they present them with privacy policies. To the extent that the transcripts contain the actual language used, analyses of individuals' speech styles combined with what they do in the app could lead marketers to draw conclusions that affect, for example, whether they get special deals. The marketers could also create data profiles about them that they share with other firms.

Is this sort of voice profiling happening today? It's impossible to know, because the back-end work of firms with voice apps is closed to public view. This lack of certainty about voice surveillance is especially concerning because the voice intelligence industry is showing up everywhere. Whereas two decades ago the notion of providing one's voice for computer analysis was the stuff of science fiction, today huge numbers of people—tens of millions and growing in the United States alone—are in the habit of speaking with virtual assistants on multiple devices. Voice technology now permeates virtually every important area of personal interaction. Car companies, home builders, hotels, and even schools are facilitators for these devices, and the devices themselves are facilitators for the voice apps within them. Watches and wireless earbuds can also send people's voices to various companies (typically via phones), depending on their creators' creativity and the marketers' desires. So far, your refrigerator isn't telling Google about your mood today, but you can't be sure that your phone isn't, that the transcript of your command to a Google Nest device isn't—or that your interaction with the voice app on your smart speaker isn't telling some other firm or firms.

People may be wary of voice surveillance, but they still are being seduced by the voice assistants' low cost and ease of use. The "stupid speakers," as some call them, may have their kinks to iron out, but press accounts—even those that politely mention privacy problems—praise their cutting-edge modernity and appeal, and for many people their exciting, convenient features are enough to quell any concerns about surveillance.

Taken together, however, these developments in voice tech complicate the very idea of personal freedom in modern society.

Voice technologies are sold under the rubrics of increased choice and autonomy—the very definition of contemporary individual sovereignty. Yet buying into voice-intelligence technologies ironically requires giving up a rich trove of information about one's body, feelings, and behavior to companies that want to use it to audit and control your behavior.

While voice-biometric technology has crept across many areas of life, the industry has managed to keep most of its activities secret. Several industry practitioners I spoke to for this book said that there is far more analysis of what people say and how they say it than companies let on. They and others also suggest that we are in a relatively dormant period. This is by design. Google, Amazon, some of the car manufacturers, and shopper-marketing agencies such as Mars are waiting for the "scale" of voice-first devices to grow until voice assistants are integrated into virtually everyone's routines.[5] Then the firms will shift into high gear: people will routinely get personal buying suggestions, search results, map destinations, and ads based on what firms conclude about them through a combination of data points including speech, demographics, behavior, psychographics, and location— all integrated into what we might call Voice+ profiles.

Merchants, who are focused on the advertising side, are virtually unanimous in predicting that Amazon, Google, and other voice intelligence providers will eventually ramp up the flow of commercial messages through their voice systems. It's not hard to speculate about the forms these will take: traditional audio ads that resemble radio ads; spoken suggestions that dovetail with voice-search queries; audio discounts (with visual counterparts in the device's mobile app) that mesh with or counter the voice search result; and audiovisual ads and discounts on voice-first devices with screens such as the Echo Display, Google's Nest Hub, Facebook's Portal, or even Samsung refrigerators.

As we saw, advertisers worry that they will have far fewer opportunities to reach customers if voice firms allow only one or two ads to dovetail with the speaker's voice-search activities. Another concern is that regulators' discomfort with data capture for advertising may cause Amazon, Google, Facebook, Samsung, Apple, and other voice-first players to delay widespread advertising via their assistants longer than the ad industry would like. For these reasons, Voice+ profiling may first show up in product suggestions from smaller firms that own their own assistants, a step that will only speed people along the path to becoming familiar— and comfortable—with having their voices used for commerce.

Eventually, improvements in voice-related technologies and even large companies' fear of getting rolled by the competition will outweigh the fear of public or regulatory disapproval, espe-cially as smaller firms and firms in sectors out of the public gaze get away with aggressively judging people by their voices, and as those small firms potentially sell the voiceprints they gather to data traders, leading people's identifiable voices to be widely distributed. At that point, personalization may go into overdrive. Researchers at MIT and Carnegie Mellon are today working on ways to reconstruct rough likenesses of people's faces based on only short snippets of their audio.[6] Marketing consultants will likely argue that demographic data taken from voice is more accurate, more current, and ultimately more cost-effective than similar, and possibly more secretly mined, data purchased from third-party vendors. Moreover, the argument may go, companies can implement voice data in real time; decisions about an individual based on gender or illness can be applied in the moment, as she or he is conversing with the voice app. All this will open the floodgates to voice analysis. A voice assistant may give an angry person greater discounts than a nervous person, even if the firm wants to keep both customers. An assistant may treat a woman whose voice lifts at the end of her sentences differently from one who speaks in ways it considers less tentative. Or a person who "sounds" unsophisticated may not get the level of explanation, or the offers, extended to one perceived to have higher status.

It's not just the opportunities for discrimination that should worry us. Smart speakers have become the epicenter of conversational interactions between individuals and thermostats, lights, and other smart-home devices in our most private spaces. Beyond any specific persuasive message people may receive, the habituation that is being encouraged by Amazon, Google, and Apple, together with Samsung (general appliances), Philips (lighting), Lutron (smart light switches), and other manufacturers, is creating a new hidden curriculum for American society. That is, together these companies are teaching people that giving up your voice is part of getting along in the twenty-first century. Educational sociologist Philip Jackson coined the phrase *hidden curriculum* during the 1960s, but the idea goes back much earlier.[7] The sociologist Émile Durkheim wrote his lectures on *Moral Education* (published posthumously in 1925) about the generalizations that students inevitably stitch together as an unspoken behavioral code. Jackson followed Durkheim in exploring the under-the-hood concepts that children pick up as part of their schooling. Other sociologists describe the ways that schools present students with "an approach to living and an attitude toward learning; how they implicitly lay out norms and values that are crucial for navigating the outside world; and how they inculcate in students the structures of social power and their relationships toward them."[8] Communication scholar George Gerbner convincingly generalizes this educational process to the media, arguing that "culture powers" from business, government, education, medicine, the military, and other areas of society deeply influence widely shared views of reality. The media's power to affect this hidden curriculum, he writes, gives it "the ability to define the rules of the game of life that most members of a society take for granted."[9]

It's no great stretch to see how the idea of a hidden curriculum can apply to the voice-marketing relationships that people learn to consider normal. It also makes sense that once people accept the concept of giving up their voice to marketers, they will accept it in other key areas of contemporary life—such as political campaigns, immigration and asylum-seeking, and prisons. Consider campaigns. Political candidates have long wanted to pursue personalized advertising to the greatest extent possible. As early as 1892, according to historian Michael McGerr, Republican National Committee chair James Clarkson boasted that he had, "with two years of hard work, secured a list of the names of all the voters in all the important States of the North, in 20 or more states, and lists with the age, occupation, nativity, residence and all other facts of each voters' life, and had them arranged alphabetically, so that literature could be sent constantly to each voter directly, dealing with every public question and issue from the standpoint of his personal interest."[10] The rise of mass media dampened enthusiasm for individual targeting during the first half of the twentieth century, but by the early 1960s, the use of market segmentation in commercial advertising was influencing political marketing. In his 1960 primary campaign, John F. Kennedy collected large amounts of data about voters' opinions and values, and used it to refine his message for different audiences and transform himself from a relative unknown to his party's nominee for president. Political campaigners increasingly employed pollsters to help them identify messages that would resonate with various voter segments. These initiatives drew on the development of "psychographic" marketing, which relied on a combination of demographic and psychological information to create homogeneous market segments.

In the early 2000s, campaigns began to adapt techniques from commercial advertising, analyzing masses of consumer data in order to predict the behavior of individual voters. Among the first to use the technique was Mitt Romney in his successful 2002 run for governor of Massachusetts. Using a tactic known as microtargeting, a Romney consultant named Alexander Gage found and combined information about individuals' political preferences and consumer habits. These were then added to the Republican Party's comprehensive database of information on voters so that these individuals could be targeted—usually by the traditional avenues of phone and direct mail—with messages designed specifically for their groups.[11]

The spread of the web and mobile phones during the 2000s transformed those practices in three key ways. One was a campaign's unprecedented ability to gather enormous amounts of information about individuals by getting them to register on websites, purchasing information about them, following their activities on the web, and noting the locations of their digital devices—their desktop computers, laptops, tablets, mobile phones, even gaming consoles. Another game changer was the use of sophisticated computer models that process enormous amounts of data to identify the most and least desirable individuals and groups from the standpoint of a particular campaign strategy. The third was the ability to reach people through a variety of digital platforms—advertising on websites, ads on Google and Bing search engines, email, social media such as Facebook and Twitter, and more—at the very moment a campaign believes such pinpointing will be useful.

It's clear, then, that political campaigns used personal data long before Cambridge Analytica caused a furor by poaching personal Facebook data in order to tailor Trump election ads sent to millions of Americans in 2016. In view of election campaigns' desire for rapidly available data and the enormous amounts of money they need to spend in short periods of time, it would hardly be surprising if they latched on to the spiral of personalization and rode it as far as possible. It is easy to imagine political phone campaigns, for example, analyzing individuals' voices in real time (much as some contact centers do today) to infer whether people are, say, angry, frustrated, or optimistic in relation to certain topics, and then having a computer feed the campaign representative the most persuasive responses for that scenario.

Some campaign leaders might argue that having people's voiceprints on file is useful because it lets campaign reps know what to expect in advance of calling or visiting voters. Look for political operatives in future years to advocate that people's voices should even be part of states' voter files, to be used as proof of identity and registration. Many states might keep voice-prints confidential, as they do with other personally identifiable information. But other states, like Maine and Nebraska, allow voters' confidential information to be used as long as the use is "noncommercial."[12] Such permission, of course, would allow campaigns relatively easy access to voice profiling, with potentially profound consequences: political operatives might use complex data analyses linking individual voters' voices to their demographic, behavioral, and psychographic characteristics to help persuade those voters across a variety of media and in-person events.

Another high-stakes area where voice intelligence seems ripe to intervene relates to asylum seekers. Basic biometrics has already made its way into the process via the United Nations. Many individuals escaping horrors in their home country contact the United Nations High Commissioner on Refugees (UNHCR) in the place to which they have fled. UNHCR personnel will then enter their data into their case management systems. If UN personnel suggest that the refugee should seek asylum in the United States, the UNHCR sends an electronic file to U.S. Citizenship and Immigration Services, part of the Department of Homeland Security (DHS). In 2019, UNHCR entered an agreement with U.S. Citizenship and Immigration Services to routinely share certain biometric data—fingerprints, iris scans, and face images—on the people it refers for resettlement in the United States.[13] In that way DHS began building biometric profiles on individuals who apply for asylum. The agency states that the biometrics from UNHCR "are used to

help establish an initial identity" that will help immigration officials "prepare for and inform the interview process, and to screen
and vet the referred individual," including checking "for previous
encounters and identities that might contain derogatory information on an individual." This information is placed along with
demographic and biographic information in a DHS repository
called IDENT/HART. "Once an identity has been enrolled in
IDENT/HART," according to the agency, any "additional collections of biometrics and biographies linked to that enrollment" will
be gathered there.[14]

As of 2019, DHS did not include voice profiling among the
biometrics it collected. One can easily imagine, though, that under
political pressure for fast and "objective" decisions about who
does or does not belong in the U.S. (especially at the southern
border), immigration officials might try to infer age and country
of origin via computer speech analyses as they work to determine
whether a person has given a truthful biography. They may also
draw conclusions about emotions and personality from the way
that asylum seekers tell their stories. Obviously, such conclusions
can be fraught with controversy. Voice researchers claim that the
rules for analyzing voice profiles apply to everyone, regardless of
a person's debilitated state or country of origin, but these generalizations haven't been tested. Moreover, the pressure to create
quantitative indexes that seem more credible than a border agent's
intuition may override concerns about using unverified voice
profiling among such highly stressed populations.[15]

Commercial marketing, voting in public elections, and
seeking asylum are critical processes that help decide what goods
individuals can access in the society and at what price; who will

assume political leadership; and who can belong to the society and participate in the commercial and political spheres. But of course, voice bio-profiling doesn't have to stop (or even start) with these three arenas. Once the power to judge people by their speech has been demonstrated in commercial or political marketing, it's likely to be picked up by just about any power center with the creativity to exploit it. Every social institution—police, military, schools, the health system, the legal system—will find ways to make claims about people based on how they speak. We can expect that the claims will run far ahead of the science, and the science will run far ahead of the ethics.

In New York, Texas, Florida, and other states, officials are paying Securus Technologies and Global Tel Link to extract and digitize the voices of incarcerated individuals and the people they speak with over the phone. The technology was created for the U.S. Department of Defense in order "to identify terrorist calls out of the millions of calls made to and from the United States every day." Through those firms, prison authorities have brought hundreds of thousands of voiceprints into large-scale biometric databases. Securus and Global Tel Link algorithms then analyze the databases to identify the individuals taking part in a call, and to map the networks of calls the individuals made. In some jurisdictions—New York is one—the software analyzes the voices of call recipients "to track which outsiders speak to multiple prisoners regularly." Wardens see this biometric identification as a way to be sure that incarcerated people don't make calls using other prisoners' PINs and that they are not speaking to former prisoners in patterns that suggest illegal activities. But rights advocates worry that people who were never incarcerated (such as family members) are having their voiceprints ensnared, and they point out that state corrections agencies seldom mention the voiceprint databases to inmates or their families. Jerome Greco, a forensics attorney at New York's Legal Aid Society, told journalists from *The Intercept* and *The Appeal* that he could understand the value of certain monitoring activities, since incarcerated individuals who want to make more calls than allowed sometimes take the phone PINs of others. Yet he worried about the monitoring of people outside prisons—he believed the prisons should have to get a warrant to do that—as well as the lack of regulation or transparency regarding how the thousands of files could be used. Voiceprints shared with police, for example, could be compared against voices caught on a wiretap, despite scientists' skepticism about the reliability of voiceprint matches for criminal prosecutions. "Once the data exists," he said, "and it becomes part of what's happening, it's very hard to protect it or limit its use in the future."[16]

How do commercial marketers, campaign marketers, DHS, and state prison systems get away with analyzing, using, and storing voices without individuals' permission? In general, when an arm of the government uses citizens' personally identifiable information (including biometrics) without the person's permission, it must be specifically required by law (think of tax returns), a judge must allow it through a subpoena, or the data must be collected in public (police can gather voiceprints from street conversations, for example). Asylum seekers, however, aren't U.S. citizens, and prisoners sometimes are treated as if they aren't. In the prisoners' case, states continued to collect voiceprints from their conversations despite complaints by prisoner-rights advocates that, at the very least, collecting prints of the non-incarcerated people on the other end of the line requires court permission.[17]

By law, asylum seekers are to be given more information than people in prison about the collection of their biometric information, but the law is seldom followed. The Homeland Security Act of 2002 states that a DHS chief privacy officer must ensure that personally identifiable information "is handled in full compliance with the fair information practices set out in the Privacy Act of 1974." Those principles are

> **transparency:** notify the individual of the data use
>
> **individual participation:** "to the extent possible, seek individual consent"
>
> **purpose specification:** state the purpose for which the information is to be used
>
> **data minimization:** collect only the information needed for the stated tasks, and retain it for only as long as necessary to accomplish those tasks
>
> **use limitation:** use the information only for the purpose collected
>
> **data quality and integrity:** ensure that the information is "accurate, relevant, timely, and complete"
>
> **security:** safeguard data against loss and unauthorized access, and
>
> **accountability and auditing:** train employees and contractors to comply with the principles.

In practice, the "private impact assessment" based on these principles lays out the extent to which DHS is able to comply with them. Sometimes the agency cannot evaluate the accuracy of UNHCR's initial data, and it cannot ensure that asylum seekers will understand or even be able to consent to the husbanding of biometric data about them. These "risks" of complying with the principles have not, however, stopped DHS from collecting and using the data to assess asylum seekers.

While government surveillance experts sometimes have to go through hoops to get permission for, or justify, their biometric poaching, the right of commercial and political campaign marketers to collect and implement voice data about individuals is often more straightforward. That's because apart from statutes governing certain areas, such as health, financial information, and children under thirteen years old, no federal law exists to restrict merchants' surveillance practices. The Federal Trade Commission merely requires that firms follow privacy policies that govern how they use people's information, and that they make these policies publicly available. A U.S. company can generally use and keep any shopping or other commercial information about individuals without permission as long as its privacy policy describes, even vaguely, what the firm does. These minimal restrictions also apply to voiceprints and voice transcriptions.

There have been some attempts at change. With increasing complaints by advocates, the press, and public officials over the streams of individual data that companies are using, the past few years have seen movements for restraints on free-for-all data collection. Congressional disagreements stalled a federal commercial-privacy law, but a few states have stepped into the breach.

The most high-profile state law is the California Consumer Privacy Act (CCPA), which went into effect in 2020. It defines biometrics as "an individual's physiological, biological or behavioral characteristics, including an individual's deoxyribonucleic acid (DNA), that can be used, singly or in combination with each other or with other identifying data, to establish individual identity." The law then goes on to list many biometric identifiers, including a voiceprint. California doesn't require that a company get a person's permission to gather and use a voiceprint. It's acceptable to provide notice "at or before the point of collection"—for example, in a privacy policy. But a company must delete the voiceprint if the individual asks. In addition to giving people various rights over the information the firm gathers, any company that traffics in people's personal information (including voiceprints) must place a prominent opt-out link on its website that says "Do not sell my data."[18]

Nothing in Bank of America's or Pandora's current policies prevents the companies from using voice profiling in California with the CCPA in place. Pandora's privacy policy states explicitly, "If you grant microphone access to the Pandora app on your device, we will receive your voice data when you interact with a voice feature on our app."[19] A separate FAQ states that Pandora and SoundHound (which creates the assistant) may "also use voice data for purposes set forth in their respective privacy policies, including . . . to optimize the product and user experience"—which can mean anything the companies want, including voice profiling.[20] But those cryptic comments about surveillance are far down on a web page that exhorts Pandora users to "Just say, 'Hey, Pandora!'" in order to get music, podcasts, and playlists.

Speaking to Pandora, it adds, is "awesome for driving, great for chores, perfect for parties"—that is, at home and outside. If users talk while they listen, Pandora will be able, based on its privacy notice, to link individuals' voice profiles to a host of facts collected about them, including their locations. These, in turn, can enable discriminatory approaches that vary ads, discounts, and offers based on the conclusions that Pandora and SoundHound draw about them.

Bank of America also gives itself room to use its customers' voices for tailored marketing. The guiding federal law for financial institutions has a floor of data requirements but a weak ceiling, especially if customers agree to the use of their data.[21] Bank of America's "U.S. Online Privacy Notice" states that it uses "personal information" to "deliver marketing communications that we believe may be of interest to you, including ads or offers tailored to you." Like other firms deep into machine learning, the bank comments in an FAQ that it keeps recordings of a person's "discussions" with its voice assistant, Erica, for ninety days to help "refine listening skills." Among the reasons it records a person's conversations, it notes, is to "identify opportunities to make Erica's responses more helpful and ensure Erica's performance is optimal."[22] The goal, according to the firm, is to fulfill a "full range of personalized, proactive insights." It may be that at present Bank of America is refining listening skills for basic language accuracy rather than anything to do with using machine learning to draw conclusions about a person from that individual's voice. Down the line, though, "optimal" performance could flow from a belief that voice personalization matters. The idea is certainly circulating in the banking industry. As early as 2018, a banking-oriented website reported that 78 percent of people asked by the voice analytics firm Invoca "said they'd use their voice assistant more often if it could better understand their tone of voice." It added that a report from Capgemini consulting concluded that the voice interface "promises to be a curator of services and experiences that intelligently meet needs and engage consumers emotionally—anytime, anywhere."[23]

Bank of America's online privacy notice doesn't exclude such activities. To the contrary, it mentions "voice to text queries" as "among the personal information we collect online." It then states that "personal information collected from and about you online described in this Notice may be used for many purposes." That could mean using what the Bank learns about you through voice to "personaliz[e] your digital and mobile experience." Another purpose for personal information like a voice profile is for "advertising on our Sites and Mobile Apps as well as non-affiliated third party sites and through off-line channels like financial centers, call centers and direct marketing (for example email, mail and phone)." The notice adds that the company reserves the right to perform "analytics concerning your use of our online services," which could involve what you say to Erica, how you speak to it, when, and where.[24]

While it's clear that the Pandora and Bank of America apps, given their privacy policies, can obtain data from customer voices in California, it's far from clear that they could do it in Illinois, Texas, or Washington, all of which have restrictive laws specifically around biometrics. The 2008 Illinois Biometric Information Privacy Act (BIPA) set the pattern for the others and is the only state law allowing private suits and recovery damages for violations. (In the other states, the government must take the initiative to file suit.) BIPA describes a *biometric identifier* more narrowly than the CCPA; it's "a retina or iris scan, fingerprint, voiceprint, or scan of hand or face geometry." Within this narrower orbit, BIPA takes an explicitly opt-in approach to companies' use of biometric data, including voiceprints. The statute prohibits "private entities" from gathering a person's biometric identifier unless the entities inform that person or the person's authorized representative of the collection, state the collection's purpose and length of time the identifier will be held, and get a "written release" from the individual. The law further notes that "no private entity in possession of a biometric identifier or biometric information may sell, lease, trade, or otherwise profit from a person's or a customer's biometric identifier or biometric information."[25] The fines are set by the law—$1,000 per violation, and $5,000 per infraction if the party has intentionally or recklessly violated the act. The total can add up to huge amounts because voice firms often deal with very large numbers of people. Moreover, the Illinois Supreme Court ruled that the act requires individuals only to show that a company violated the law—not that they were harmed—in order to file a lawsuit.[26]

This Illinois BIPA requirement for a written release led to a lawsuit against Amazon that was mentioned in Chapter 4. Amazon's defense is that the terms of service for Alexa satisfy BIPA's requirement that users be notified in writing regarding the company's policies for using and storing biometric information. As of September 2020 it had asked the court to send the claims to arbitration if it won't dismiss them. So far, no BIPA class actions have targeted Google or Apple for their voice activities, though it wouldn't be surprising, especially if Amazon loses or settles. A Google public relations representative told me that users of the Google Home are asked to opt-in to the company's policy of recording the user's voice when they set up the device's app; they can also set the amount of time the company will keep the information. As I noted in Chapter 3, actual written releases are already the practice in the automotive industry, where new owners sign statements of terms and conditions for their car's voice assistant when they sign other purchase documents. Most states, even those following the California privacy law, don't even require a push-button opt-in (for example, pressing "OK" in response to information about what data are collected and why). But while only the Illinois law is at the center of the litigation, the Amazon lawsuit may eventually force the voice intelligence industry to confront a basic proposition nationwide: the least burdensome way to gather voiceprints without litigation will be to encourage individuals to give opt-in permission (written or not) for their voice and speech to be used for personalized, targeted presentations.

It may not be difficult to get this permission for commercial purposes. It is not at all hard to imagine that people would hit an OK button to allow a firm they know to give them news, views, and tools tailored to their own lives. I've already discussed in detail one reason for this: companies and the mainstream media have spent several years habituating Americans to share their voices in even the most intimate spaces. It's become second nature for many to ask Google for directions, Alexa for a recipe, and Siri for the weather in a city they will be in the next day. People with connected thermostats and cameras instruct the devices through their smart speakers and phones. Over time, a generation that has grown up with personal assistants will take it for granted that those assistants are listening all the time. And for those who need an extra push—and whose lifestyles are worth monetizing and tracking—companies may well make offers tantamount to bribes. To some extent, that already happens. Amazon and Google have priced some Echo and Google Home products so low that it's clear they are priming the pump for revenues that will eventually come through targeted advertising and the sale of products.

As people continue to use these devices, not only do voice-first technologies become a more familiar, comfortable part of their daily lives; individuals also become resigned to the surveillance part of the experience, feeling that little can be done to stop it. Companies encourage this attitude toward all sorts of digital practices. Their privacy policies, written in the jargon of lawyers, are hard to follow and puzzlingly vague. As it turns out, very few people even try to read them, and one reason may be the label itself. National surveys have consistently found that when given a statement such as "When a website has a privacy policy, it means the site won't share your information without your permission," more than half of the respondents don't know that the statement is false.[27] Neither this privacy policy research nor the work on resignation mentioned voice profiling or any other form of biometrics. Yet the strategy of seductive surveillance that companies use to inculcate the voice-assistant habit fits hand-and-glove with the larger corporate mission of making surveillance an accepted part of life. As the voice intelligence industry grows more widespread and confident, it will trumpet its ability to bring new levels of real-time personalization as a necessary and desirable extension of traditional marketing strategies for twenty-first-century users. In time, the unrelenting urge to find out ever more about customers and potential customers—as well as voting targets and others—will lead to the use of additional bio-profiling elements. Perhaps faces will join voices in telling tales about people in their homes and elsewhere. Perhaps the spiral of personalization will lead to other biometric storytellers—heartbeats? urine?—that can give marketers new advantages. Politicians and government agencies will follow; sometimes they will lead. And individuals will continue to opt in, giving up their data, feeling they are getting a good deal in the moment, vaguely worried that bad things may result, but resigned to the transaction.

And instead of working to ease inequalities based on income, race, ethnicity, and education, the roiling competitions among commercial, political, and government interests will drive social divisiveness forward. These effects are not new, nor unique to biometrics. Marketers have been breaking up America conceptually and practically since the mid-twentieth century.[28] In recent years they have been picking individuals apart from one another, treating them differently based on their demographic and psychographic characteristics. But the use of the body to further separate people is a line no good society should cross. While companies may boast that they have honed personalization to a science, it's crucial to realize that the very nature of the training sets used to improve voice AI helps set the stage for an important form of discrimination. As we have seen, if voices belonging to people of color, of certain ethnicities, with certain accents, or with certain speech irregularities don't make up a large enough population in the computer training, their voiceprints may be counted as irregular and not deserving attention—or they may be associated with narrow descriptions of people (good or bad) based on the small sample. This effect is known as algorithmic bias. If the baseline values of the voice profiles are set by samples that are overwhelmingly white, American, English-speaking, and middle-class (because that's who has the purchasing power marketers want, and the high-speed internet access to use these devices in the first place), it seems at least possible that voice tech will misinterpret the emotional and personality signals of people who are poorer, nonwhite, or from other cultures. It is also quite possible that bio-profiling in commerce and elsewhere will push people further into the silos of deals, news, views, relationships, and life expectations linked to allegedly hardwired attributes those people may not fully understand and, because they are physiological, may not be able to escape. Perhaps marketers' voice inferences about customers will change as the customers age. But should people have to rely on that to change their profiles, and would they always want age to be a determining factor? Such unanticipated consequences of bio-profiling, once most people have bought into it, will lead to less individual flexibility and freedom and, in turn, more social fragmentation and distress.

So what can be done? Does any nation present a good working model for regulating these ever more aggressive voice-intelligence technologies? Here the examples of China and the European Union may be instructive. China resembles the United States in having sped ahead with voice intelligence, and its approach to all sorts of digital marketing, including for smart speakers, the connected home, and related activities, offers a tantalizing if ultimately unsuc-cessful guide. That's unfortunate because many people in China seem to recognize the problems that marketers' bio-profiling can bring. When it comes to most of its voice intelligence industry, and especially its smart speaker market, says voice-marketing expert Bret Kinsella, China is mostly self-contained. "With the exception of Apple," Kinsella wrote in 2019, "none of the key non-Chinese smart-speaker makers sell in China today and the presence of Chinese smart speakers outside the country is minimal."[29] There is enormous competition among the leading companies, chiefly Baidu, Alibaba, and Xiaomi. Baidu, the country's dominant search engine, was in 2019 the dominant player in phones, autos, and smart speakers, according to research firm Canalys. Clearly wanting to be "everywhere," Baidu announced in 2019 that its DuerOS voice assistant could be accessed on 400 million devices.[30] In 2019 its smart speaker zoomed past Xiaomi's and beat that of the previous leader, Alibaba, partly through an aggressive low-price strategy much like Amazon's and Google's in the United States.[31]

This competition and technological change are dovetailing with important alterations in marketers' surveillance policies. Until around 2018, Baidu, Alibaba, Tencent, Xiaomi, and smaller digital marketers had free rein to gather their target audiences' data, including voice data, in the name of personalized service. "Chinese companies are increasingly finding that the days of collecting data without public scrutiny are over," wrote Samm Sacks and Lorand Laskai in *Slate* in early 2019. "Chinese consumers are vocally standing up for their own privacy in ways not seen before."[32] The national government seems to be behind this change. When Baidu's founder suggested in an interview that the Chinese people would trade privacy for convenience, Chinese state media reported widely on users' anger with the sentiment. In January 2018, China's Ministry of Industry and Information Technology warned Baidu that it had to increase protection of its users' personal data.[33]

Sacks and Laskai described an emerging Chinese data protection system "with rules for consent; personal data collection, use, and sharing; and user-requested deletion of data."[34] Called the Personal Information Security Specification, the guidelines took effect in May 2019 and were not legally binding, though government regulators have been pushing marketers to recognize them as if they were law. The strong emphasis is not just on being up-front about the data collected for marketing, but also on limiting pressure on individuals to agree to what regulators call "the excessive collection of personal information." To show that the government was serious, in 2019 the Ministry of Industry and Information Technology blacklisted fourteen apps it claimed had "excessively collected sensitive personal data" without user consent.[35] The same year, four powerful government ministries used provisions of the existing Cyber Security Law and Consumer Rights Protection Law to evaluate nearly a thousand apps in ten topic areas for "compulsory and excessive collection of personal information."[36] The aim was to encourage developers to clean up their apps in advance of the public judgments.

Publicity about these sorts of actions led observers to herald China as a leader on data privacy. Sacks and Laskai wrote that "given federal inaction in the United States on consumer data protection, on paper, at least, Chinese consumers might soon have greater privacy from tech companies than American consumers."

Unfortunately, though, it doesn't appear that the Chinese regulators have defined "excessive" and "coercive" in ways that can serve as reliable guides to how sites and apps should proceed. Moreover, Beijing's strong pull for near-total control over the population's digital world has created an environment in which companies still must create and store data for government use without individuals' permission, even as they are supposed to be transparent and get permission to use the data in marketing. "For example," write Sacks and Laskai, "China's e-commerce law requires companies to delete user data but also mandates that companies retain data to assist with government investigations for national security. China's cybersecurity law requires consent to collect personal information, but it also grants the government new powers to demand that companies turn over more information on users through random inspections of internet service providers, making it increasingly difficult for users to be anonymous online."[37] All this led observers to question whether the government will really prosecute companies that use data they collect surreptitiously for the government.[38]

Oddly, the U.S. approach to commercial surveillance may be more similar to China's than to the European Union's. That's because the European Union philosophically considers privacy in this context to be part of people's "fundamental rights and freedoms."[39] Central to the European Union's approach to commercial privacy are two acts, the General Data Protection Regulation (GDPR), implemented in 2018, and the ePrivacy Regulation (which replaces an earlier ePrivacy Directive), which will likely go into effect in the early 2020s, after much wrangling by EU representatives over certain features of it.[40] The GDPR lays out the basic scope of the European Union's approach to privacy, and the ePrivacy Regulation (which is much like the ePrivacy Directive that will stay in effect until it passes) applies those ideas to any business that "provides any form of online communication service, uses online tracking technologies, or engages in direct marketing."[41] The GDPR defines biometric data as "personal data resulting from specific technical processing relating to the physical, physiological or behavioural characteristics of a natural person, which allow or confirm the unique identification of that natural person, such as facial images or dactyloscopic [fingerprint] data."[42] This definition clearly seems to include voice bio-profiling, though not the use of voice transcripts. The law goes on to prohibit the use of "biometric data for the purpose of uniquely identifying a natural person."[43] Right away, though, the GDPR grants exceptions, including if an individual has "given explicit consent."[44]

Note the word *explicit*. Consent, according the GDPR, is not opt-out as it is in most of the United States. Central to the law is a belief that individuals should approve the data that a company is using about them. As a European expert notes, "the bar for valid consent has been raised much higher under the GDPR. Consents must be freely given, specific, informed, and unambiguous; tacit consent would no longer be enough."[45] And when it comes to voice, Tim Mackey, principal security strategist for technology firm Synopsis, described the regulation's approach this way: "Under GDPR, there's, for practical purposes, six possible scenarios under which pieces of data should be collected and potentially acted on. And under two of them, it's either to perform the valid function of the device, or [with] user consent.

And with user consent, there has to be a level of disclosure. What are you collecting? Why are you collecting? How long is it going to be retained for and who's going to be touching this data?"[46] Fines for disobeying GDPR or the forthcoming ePrivacy Regulation can be huge—up to 20 million euros or as much as 4 percent of a company's total worldwide annual revenues, whichever is higher.[47] Most people connected to the European Union's privacy regimes concede that the extraordinarily high penalties will not kick in until ambiguities within the laws and contradictions between them have been resolved, which may involve the courts. At present, for example, many EU websites and apps are getting away with flimsy notices about tracking that simply urge a person to hit an OK button, and instead of a decline option, they present an alternative that, frustratingly, says "learn more." It's also not clear what information Amazon, Google, and Apple are giving voice-assistant users in return for "informed" permission to record and store their voices.

Recent EU rulings related to phone recordings show that regulators are beginning to look at the direct marketers and customer service centers. In early 2019, for example, the Danish privacy regulator banned the country's largest telecommunications firm from recording customers' calls until it had offered a way to give active consent. The regulator didn't fine the firm because it was a first-time violation, but "practitioners and legal experts" told Bloomberg Law that the ruling "should serve as a warning to other businesses across the European Union that record customers' phone calls."[48] The article didn't specifically mention the European contact-center business, but its executives surely received the message.

This is not the place to detail the ways in which the GDPR and ePrivacy regulations are carried out—and fines meted out—throughout the member countries. It does seem clear, however, that the European Union's general approach can provide some wayfinders for U.S. policymakers hoping to develop sound regulations concerning voice biometrics. One strategy worth emulating is to require a specific opt-in for voice profiling, on a multistate basis and backed by the threat of huge fines. Another is to grant the courts the authority to vary those fines based on specific circumstances so that companies can sometimes only be warned. Another useful facet of the GDPR is that the individual right to consent may be disallowed in specific cases by the European Union or a member state.[49] That is, there may be some types of data collection that should never be allowed because regulators consider them so egregious, or they don't believe people know enough to give informed consent.

On the negative side, the GDPR gives too much leeway to state governments' use of personal data, allowing member states to adopt special codes that circumvent parts of the regulation. The European Union also typically relegates decision-making about national security and crime to its member states. As Chris Hoofnagle and colleagues note, the EU's Police Directive, which lays out a framework for data processing by law enforcement agencies, has rules that "allow for more limitations than the general framework provided by the GDPR."[50] As the border and prison examples presented earlier indicate—and as China's notoriously oppressive use of biometric and other personal data reveals—it's important to be at least as wary of law enforcement as of marketers. Yet another drawback of the GDPR and eDirective is the omission of voice transcripts. As companies learn how to apply highly sophisticated analyses of individuals' word use and speech patterns—even including pauses and stutters that could appear in transcripts—they may ramp up voice-based profiles to enhance personalization models. Recall from Chapter 2 that this is already happening in the contact-center business. It's quite likely that if bio-discriminatory activities based on voice stall for a while due to public concerns, linguistic strategies will pick up the slack—and these are similarly dangerous in the sense of pushing personalization in ways that lead to social dysfunction.

The presence of the GDPR and the laws in California, Illinois, Texas, and Washington State are fueling interest by policymakers in several states for laws that will rein in biometric surveillance.[51] The state leaders feel compelled to act in large part due to the lack of strong federal action. As Lauren Bass wrote in a 2019 law review article, Congress, "afraid of stifling innovation and dampening the progress of capitalism . . . to date has maintained an arms-length approach to regulating both Silicon Valley and its Big Data practices." She added that "as AI voice-technology continues to infiltrate and embed itself into the daily fabric, the prevailing U.S. legal framework . . . is ill equipped to efficiently or effectively regulate the predatory privacy procedures of data-hungry tech companies, leaving consumers prey to abuses and violations in the collection, storage, and manipulation of their personal information."[52]

This hands-off approach seems especially dangerous when it comes to voice and speech profiling. Because voice is an intimate and unchangeable individual feature, the surveillance and profiling that accompany this new level of personalization should be held to a higher standard than past forms of targeting, one that will set a high public-interest bar for all the bio-profiling forays that will follow. We've reached a point where a line must be drawn, beyond which marketers should not be allowed to go. If the use of voice profiling in the most private areas of a person's life goes unchallenged, companies will follow the unending spiral of personalization to profile us based on our faces, hair, blood, and any other aspects of our anatomy and physiology that can send signals about us. It's not that far-fetched to imagine a device that can monitor and decipher the electrical activity of our brains. If these speculations seem remote or outlandish, consider whether thirty years ago people would have thought it plausible that their telephones would track them whenever they leave the house, that the stores where they shop would use the same phones to trace their movements in the aisles, and that they would ever allow in their bedrooms a cylinder that they knew might record their conversations for unknown purposes. As we have seen, companies are perfecting the art of seducing people into adopting technologies that intrude into their personal spaces and lives, without fully acknowledging the surveillance taking place—and while making it difficult for users to disengage.

The case for prohibiting voice profiling is based on two related arguments. The first is that the data world of voice marketing is inherently deceptive, and thus within the U.S. Federal Trade Commission's power to outlaw it. The second is that the voice intelligence industry's intrinsic inability to obtain meaningful informed consent from individuals around its use of their bodies means that these companies' advertising chronically misrepresents their products' performance. The FTC has long had the responsibility of curbing deceptive advertising.[53] As we have seen, Amazon, Google, contact-center companies, and other voice firms have been evasive and uncommunicative regarding their key activities, aims, and strategies. The arc of deception involved in those linked activities did not start with voice profiling. It has been endemic to marketers since at least the rise of digital personalization at the turn of the twenty-first century. They have been getting away with these activities because government officials have been loath to call them deceptive. What's new with the rise of voice is that personalization is now linked to the body, and as such it needs special attention. As we have seen, legislators have begun to recognize that information extracted from the body is more sensitive, and requires more care in obtaining permission, than traditional descriptors such as demographics and lifestyle categories. What lawmakers haven't done yet, but should, is to follow the consensus of ethicists that taking information from a person's body requires meaningful informed consent. This requirement is significantly stricter than the one that simply prohibits consent obtained through outright deception.

Bioethicists, who have debated informed consent in medical and scientific research settings for decades, agree that meaningful informed consent involves two distinct steps. First, the person asking for consent should "tell potential research subjects about the relevant risks and benefits of participation, and the goal of the research." Second, whoever is asking for consent "should also allow potential participants to exercise their autonomy by accepting or refusing to participate."[54] In some cases—telling a person about specific procedures in an operation, for example—meeting these requirements is not difficult. But when it comes to other activities involving the body (getting people's permission to use their skin tissues for a variety of investigations in biobanks, for example), the risks may change over time, may be various and subtle, and may unintentionally violate the patient's values. In such circumstances, medical workers struggle to reach the required level of understanding with patients, and often wonder whether it's possible.

Such risks and concerns are compounded when it comes to the continually changing world of voice marketing—where even the companies involved sometimes don't believe their counterparts' assurances, and where the potential applications of voice-related data are still in early stages, with fortunes waiting to be made by companies racing to innovate their way more deeply into our personal lives with the harvesting, storage, and analysis of voice data. When it comes to gathering and selling data about an individual's voice, it is safe to say that no opt-in practice or privacy policy can adequately explain all the short- and long-term implications. Meaningful informed consent is therefore impossible—any representation of it is inherently deceptive.

So what are we, as American citizens, to do? What recourse do we have in the face of such challenges to the privacy of our voices, our bodies? One important step we can take now is to pressure our leaders to adopt policies and regulations that can protect these vital rights. I suggest the following priorities for U.S. policy regarding voice intelligence:

1. Voice authentication ought to be allowed in a business context if the people involved are explicitly informed of the activity and given the choice to opt out or not. This

form of biometric validation has been useful alongside other means (such as passwords) for companies like financial firms that need to confirm the identity of a person phoning about an account. Companies should, however, be prohibited from punishing individuals who don't allow voice authentication. Third-party collections for the purpose of voice authentication across companies should also be prohibited unless the third party gets explicit permission from each individual for the storage and for each use. The notion of collecting voiceprints from many companies and then having a central computer use them for authentication on behalf of other firms is troubling because the individuals do not know that a third party is storing and using their voices. Government agencies should be held to this same standard for keeping voice records of citizens and non-citizens. Use of voice authentication by a government agency without the person's knowledge should require court permission.

2. Voice identification—determining who a person is among a large population of people whose biometric characteristics have been stored—can be helpful for law enforcement but should be used only with a warrant to search a particular individual or individuals. Otherwise, authorities could use the technology to find people who haven't been charged with a crime. Voice identification should be prohibited for commercial activity. In the future, companies could use it in real time to identify a person in a crowd, say by placing microphones throughout a store (something it seems they cannot yet do), and then linking

that person to many known data points in order to tailor views and offers. No opt-in practice or privacy policy could adequately explain, to the point of meaningful informed consent, the implications of voice identification for such a target individual's situation compared to others. If a case can be made that such an explanation can be offered, the prohibition can be rethought.

3. Voice profiling, the process of drawing inferences about an individual based on his or her physical voice characteristics, can take place with biometric authentication or identification, or it can be used by itself. Contact centers, smart phones, smart speakers, and most other voice-interaction devices aim to use both authentication and bio-profiling, and sometimes linguistic assessment. The reason is clear: if analysts know the name of the person whose body they are picking apart digitally, they can use advanced analytic techniques to link the person's vocal characteristics to demographic and other categories to create deeper inferences that may help them predict what that person will do in the future. But *all* voice-profiling, linked and unlinked, should be flatly prohibited in companies' interactions with individuals. That includes the emotional profiling that is already taking place. Computer experts are only at the cusp of being able to carry out these activities, accurately or not. Nevertheless, it is quite clear that it is impossible to explain the short- and long-term implications of voice profiling for a particular individual, and that any such attempt cannot produce the meaningful informed consent that ethicists require for interrogation of the human body. Amazon's

Halo, described in the Introduction, may seem to deserve an exception because the company requires an opt-in permission to evaluate the device owner's voice, and Amazon never stores the data, which is encrypted. Yet because at this writing it's not clear where Amazon's inferences about users come from (perhaps training sets based on conversations between Alexa and people who had no clue they were contributing to the new device?), it's hard to see why this sort of voice profiling should be acceptable. The prohibition should also apply to government agencies. Individual exceptions for criminal or national security reasons should be granted only by a court with a special expertise in the conduct and consequences of biometric activities.

4. All of these limitations ought to apply as well to political campaigns and most government activities. Some political operators will argue that their essentially unlimited right to use voice authentication, identification, and profiling is protected by the First Amendment of the U.S. Constitution, which states that "Congress shall make no law . . . abridging the freedom of speech or of the press." But campaigning is marketing pure and simple, and political operatives should not be allowed to get away with activities that Americans feel are toxic to society in general and the political system in particular. If the courts don't agree, the people ought to push for a constitutional amendment that outlaws biometrics from political consulting and campaigning. The issue is that important.

5. The use of transcripts for linguistic analyses that aim to use an individual's unique speech patterns to draw

conclusions about an individual's background, activities, personality, sentiments, or emotions should be subject to the same regulations as voice bio-profiling.

Voice assistants are not our friends. They are not, as some in the media have characterized them, like the service staff at Downton Abbey. As we saw in Chapter 1, voice firms play into such myths, investing their devices with personalities so that people will want to speak with them and reveal their emotions. "When she recognizes you're frustrated with her, Alexa can now try to adjust, just like you or I would do," said Rohit Prasad, Amazon's chief Alexa scientist.[55] But buying into the friendly humanoid fallacy—and sharing that fallacy with children by calling personal assistants *she, he, him,* or *her,* rather than *it*—courts the danger of accepting the voice-exploitation future that they represent. It also sets up the problem of accepting, or leading children to accept, gendered biases that some of the assistants perpetuate. That is not only because their solicitous female voices perpetuate harmful stereotypes, but also because their mild reactions to bullying—what Mar Hicks has called "submissiveness in the face of anger"—create painfully inappropriate models of harsh speaking that children, and adults, can apply to everyday life.[56]

For more than a hundred and fifty years, Americans have eagerly yielded to virtually any techniques that marketers have developed to maximize persuasion. Any concerns brought by American citizens—for example, about internet targeting and in-store tracking—have come late, with activists having to try to reverse activities that marketing and advertising practitioners have already, with surprising speed, turned into firmly established routines. We may be facing the same situation with voice profiling. But the window of opportunity is still open. Phone assistants, though they have been around for a decade, are nevertheless rather new at voice and speech inferences. And much of our environment—our homes, our cars, our hotels, our schools—is only now being colonized by a voice intelligence industry that is still finding its footing. The instability of the industry means there may still be time to change it in important ways.

Acting now is crucial. As we have seen, companies are trying to make voice-first devices into a habit that people perceive as fun, emotionally satisfying, natural, and safe enough. But as we have also seen, the climate created by companies and the press around voice-intelligence technologies seduces—and then habituates—many to give up a part of their bodies for analysis by companies that may then manipulate them in ways they may not know, understand, or approve. We shouldn't bequeath to our heirs a twenty-first century that allows marketers, political campaigners, and governments to erode people's freedom to make choices based on the claim that bio-profiling tells who they really are, what they really believe, and what they really want. And we shouldn't base our safety on the belief that voice profiling can tell us whom we should be wary of. The voice intelligence industry may not be doing all these things yet. But as voice technologies become ever more deeply ingrained in our lives, and the incentives grow for companies to exploit what we say and how we say it, protecting those who use these technologies—in their roles as consumers, citizens, and voters, and even simply as members of their families and communities—should become a priority.

ACKNOWLEDGMENTS

To investigate the subject of this book, the new consumer surveillance business that I call the voice intelligence industry, I used a variety of research methods. I went through well over a thousand trade magazine and general news articles on the companies connected to various forms of voice profiling. I examined hundreds of pages of major federal, state, and EU laws applying to biometric surveillance. I analyzed dozens of patents. I explored the development of voice technologies over the past century to put these activities in historical perspective. And because so much about this evolving industry is not written down, I spoke to people who are working to shape it.

I owe a lot to the many marketing executives and technology experts who answered my questions about the complex technologies, business policies, and government policies that are guiding the new world. Four of these generous individuals wanted to remain anonymous. I'm grateful for the opportunity to acknowledge the others together with the organizations where they were employed when we talked: Austin Arensberg (Scrum Ventures), Dave Berger (Volara), Damian Bianchi (Global Strategies), Wally

Brill (Google), Arafel Buzan (Mindshare NeuroLab), Tony Callahan (Shea Homes), Jeremy Carney (Knobbe Martens), Gil Cohen (NICE), Tom Cowan (Knobbe Martens), Michael Dobbs (360i), Kirk Drummond (Drumroll), Scott Eller (Neuraswitch), Pete Erickson (Modev), Augustine Fou (marketing consultant), Ethan Goodman (The Mars Agency), Patrick Givens (Vayner Media), Tom Hespos (Underscore Marketing), Tom Hickman (Nationwide Marketing), David Isbitski (Amazon), Bret Kinsella (Voicebot.ai), Roger Lanctot (Strategy Analytics), Hugh Langley (The Ambient), Janet Levine (Mindshare NeuroLab), Joe Maceda (Mindshare North America), Will Margaritis (Sellwin Consulting), Hans Mayer (Knobbe Martens), Eric Montague (Nuance Communications), Peter Peng (Jetson), Joe Petro (Nuance Communications), Brandon Purcell (Forrester), Apurna Ramanathan (AskMyClass), Myklos Rambus (Equifax DDM), Brad Russell (Parks Associates), John Schalkwyk (Google), Vlad Sejnoha (Glasswing Ventures), Rita Singh (Carnegie Mellon University), Rishad Tobaccowala (Publicis), Andy Traba (Mattersight), and Brian Weiser (Group M).

In many cases I contacted sources because they either were mentioned in or wrote trade magazine or general press articles I found helpful, or because they appeared on panels at industry meetings I attended. The website Voicebot.ai was particularly useful for tracking contemporary developments in voice-related marketing developments. BiometricUpdate.com was continually helpful for monitoring the changing laws and controversies surrounding bio-physiological identification. I benefited from reports from Forrester and Gartner on the subject of voice and marketing. I am grateful for journalists (I've cited many of them) across a wide spectrum of outlets who have spent the time and energy to cover one or more of the many topics I've discussed and connected in this book. I also had the good fortune to attend the Voice 2019 conference and glean valuable insights about the emerging industry from the following panel discussants: Shlip Agarwal (Bluetag), Maria Astrinaki (Sound United), Ryan Bales (DialogTech), Tom Doyle (Aspinity), Steven Goldstein (Amplifi Media), Harish Goli (Pandora), Melissa Hamersley (Novel Effect), Mary Alice McMorrow (Earplay), Waleed Nasir (Virtual Fore), Biancha Nieves (Botsociety), Katherine Prescott (VoiceBrew), James Vlahos (writer and book author), and Charles Andrew Whatley, Sr. (Instreamatic).

Colleagues from various academic disciplines have interpersonally and in their writings been crucial to teaching me, critiquing me, and brainstorming with me about ways to think about surveillance, privacy, society, and digital media. Warm thanks are due to Frederik Zuiderveen Borgesius, Nick Couldry, Lori Cranor, Pam Dixon, Nora Draper, Jonathan Hardy, Natali Helberger, Mike Hennessy, Chris Jay Hoofnagle, James Katz, Helen Kennedy, Yph Lelkes, Tim Libert, Jessa Lingel, Lee McGuigan, Stephen Neville, Helen Nissenbaum, the late (and always illuminating) Joel Reidenberg, Tal Zarsky, and the enlightening speakers and papers at two annual meetings, the Northeast Privacy Law Scholars Conference convened by Ari Waldman of Northeastern University, and the Privacy Law Scholars Conference staged on alternate years by Chris Hoofnagle at UC Berkeley Law School and Dan Solove at George Washington University Law School.

I am fortunate to work in a world-class academic environment that provides the time and encouragement to carry out this sort of research. Two successive Annenberg School deans,

Michael Delli Carpini and John Jackson, have had a large role in creating and perpetuating that atmosphere, and I thank them for it. My thinking on the subject has also benefited from research help by Annenberg doctoral students David Cordero, Brendan Mahoney, and Andrew Wirzburger. Jenny Lee and Fallon Parfaite, also Annenberg doctoral students, helped with critiques of sense and style in my writing. Annenberg's librarian, Sharon Black, has always provided stellar and sympathetic help, as has the IT department ably piloted by Rich Cardona.

My good fortune continued at Yale University Press. Joe Calamia, the editor for my prior two books at Yale, encouraged me at this work's earliest stages. When Joe left, Bill Frucht took over and provided editorial time and sagacious advice that have left me immensely thankful. Julie Carlson continued the process with expert copy-editing that cut through verbal brush in the service of meaning and impact. Jeff Schier, who has helped me enormously on three books, was an assiduous project manager on this one. And Elizabeth Sylvia worked attentively behind the scenes to make sure all the right information and documents went to the right people. Fred Kameny produced a great index, and Robin Charney provided a punctilious proofreading. I'm grateful for all of their expertise, effort, and enthusiasm.

NOTES

INTRODUCTION

1. See, for example, Steve Olenski, "Is Voice Set to Be the Next Big Thing in Marketing?," *Forbes,* May 31, 2018, https://www.forbes.com/sites/steveolenski/2018/05/31/is-voice-set-to-be-the-next-big-thing-in-marketing/#64fd71a37d5f; and NPR and Edison Research, "The Smart Audio Report," July 18, 2018, https://www.edisonresearch.com/the-smart-audio-report-from-npr-and-edison-research-spring-2018/, both accessed October 25, 2018.

2. Dieter Bohn, "Google Assistant Will Soon Be on a Billion Devices, and Feature Phones Are Next," The Verge, January 7, 2019, https://www.theverge.com/2019/1/7/18169939/google-assistant-billion-devices-feature-phones-ces-2019, accessed December 15, 2019.

3. Ben Fox Rubin, "Amazon Sees Alexa Devices More Than Double in Just One Year," CNET, January 6, 2020, https://www.cnet.com/news/amazon-sees-alexa-devices-more-than-double-in-just-one-year/, accessed January 6, 2020.

4. Ananya Bhattacharya, "Amazon's Alexa Heard Her Name and Tried to Order Up a Ton of Dollhouses," Quartz, January 7, 2017, https://qz.com/880541/amazons-amzn-alexa-accidentally-ordered-a-ton-of-dollhouses-across-san-diego/, accessed August 24, 2020.

5. Jennings Brown, "The Amazon Alexa Eavesdropping Nightmare Came True," Gizmodo, December 20, 2018, https://gizmodo.com/the-amazon-alexa-eavesdropping-nightmare-came-true-1831231490, accessed June 3, 2020.

6. Julia Carrie Wong, "Amazon Working to Fix Alexa after Users Report Random Burst of 'Creepy' Laughter," *Guardian*, March 7, 2018, https://www.theguardian.com/technology/2018/mar/07/amazon-alexa-random-creepy-laughter-company-fixing, accessed January 31, 2019.

7. Jared Newman, "How a Family's Amazon Echo Recorded and Sent Out Their Private Conversation," *Fast Company*, May 24, 2018, https://www.fastcompany.com/40577513/family-claims-amazon-echo-recorded-their-private-conversation-sent-it-to-random-person, accessed August 24, 2020.

8. Youyou Zhou, "An Oregon Family's Encounter with Amazon Alexa Exposes the Privacy Problem of Smart Home Devices," Quartz, May 25, 2018, https://qz.com/1288743/amazon-alexa-echo-spying-on-users-raises-a-data-privacy-problem/, accessed December 5, 2019.

9. Huafeng Jin and Shuo Wang (for Amazon Technologies, Inc.), "Voice-Based Determination of Physical and Emotional Characteristics of Users," US Patent Office, October 9, 2018, https://patents.google.com/patent/US10096319B1/en, accessed August 24, 2020.

10. Jonathan Shieber and Ingrid Lunden, "Amazon Launches Amazon Pharmacy," TechCrunch, November 17, 2020, https://techcrunch.com/2020/11/17/amazon-launches-amazon-pharmacy-its-delivery-service-for-prescription-medications/?ck_subscriber_id=958959558&guccounter=1, accessed November 17, 2020.

11. Jon Arnet et al., "Synchronized Audiovisual Responses to User Requests," US Patent Application 20190180343, June 13, 2019, https://patents.google.com/patent/WO2019118364A1/en, accessed December 5, 2019.

12. Anthony Fadell, "Smart-Home Automation System That Suggests or Automatically Implements Selected Household Policies Based on Sense Observations," US Patent 10,423,135, September 24, 2019.

13. "Introducing Amazon Halo," Amazon.com, n.d., https://www.amazon.com/gp/product/B07QK955LS?pf_rd_r=D9MWFK17YZV2NVFW00V8&pf_rd_p=edaba0ee-c2fe-4124-9f5d-b31d6b1bfbee#faq, accessed August 29, 2020.

14. These examples and more will be elaborated with references in Chapter 2.

15. See Global Analytics' patent application in the United States for a "credit risk decision management system and method using voice analytics." This system "may predicts [*sic*] lending outcomes that determine if a customer might face financial difficulty in the near future and ascertains affordable credit limits for such as customer." Gopinathan et al. (for Global Analytics), "Creditor Risk Decision Management System and Method Using Voice Analytics," US Patent Application US 2015/0142446, May 21, 20015, p. 1., https://patents. google.com/patent/US20150142446A1/en, accessed July 15, 2020.

16. David Pierce, "Inside the Lab Where Amazon's Alexa Takes Over the World," *Wired,* January 8, 2019, https://www.wired.com/story/ amazon-alexa-development-kit/, accessed October 3, 2019.

17. For example, Shoshana Zuboff, *The Age of Surveillance Capitalism* (New York: Public Affairs Press, 2018); Julie Cohen, *Between Truth and Power: The Legal Constructions of Informational Capitalism* (New Haven: Yale University Press, 2019); Daniel Solove and Paul Schwartz, *Consumer Privacy and Data Protection* (Philadelphia: Walters Kluwer, 2017); Glenn Greenwald, *No Place to Hide: Edward Snowden, the NSA, and the U.S. Surveillance State* (New York: Metropolitan Books, 2014); David Lyon, *Surveillance after Snowden* (Cambridge, UK: Polity Press, 2015); Joseph Turow, *The Aisles Have Eyes: How Retailers Track Your Shopping, Strip Your Privacy, and Define Your Power* (New Haven: Yale University Press, 2017); Chris Jay Hoofnagle, *Federal Trade Commission Privacy Law and Policy* (Cambridge, UK: Cambridge University Press, 2016); and Fin Brunton and Helen Nissenbaum, *Obfuscation: A User's Guide for Privacy and Protest* (Cambridge, MA: MIT Press, 2016).

18. For example, Mara Einstein, *Black Ops Advertising: Native Ads, Branded Content, and the Covert World of the Digital Sell* (New York: OR Books, 2016); Andrew Guthrie Ferguson, *The Rise of Big Data Policing: Surveillance, Race, and the Future of Law Enforcement* (New York: NYU Press, 2017); and Virginia Eubanks, *Automating Inequality: How High-Tech Tools Profile, Police, and Punish the Poor* (New York: St. Martin's Press, 2018).

19. "Personalization," Wikipedia, n.d., https://en.wikipedia.org/wiki/ Personalization; and "Personalized Marketing," Wikipedia, n.d.,

https://en.wikipedia.org/wiki/Personalized_marketing, accessed December 15, 2019.

20. Penny Gillespie and Guneet Bharaj, "Use Personalization to Enrich Customer Experience and Drive Revenue," *Gartner*, April 3, 2019, p. 3.

21. For a more detailed history, see Turow, *The Aisles Have Eyes*, pp. 24–65.

22. For more on this period of American history, see Joseph Turow, *Breaking Up America* (Chicago: University of Chicago Press, 1996), pp. 18–36.

23. Josh Lauer, *Creditworthy* (New York: Columbia University Press, 2017), p. 7.

24. Ibid.

25. The quotation is in Joseph Turow, *The Daily You* (New Haven: Yale University Press, 2011), p. 34. For a historical overview of the period, see pp. 34–64.

26. Ibid., p. 34.

27. Safiya Umoja Noble, *Algorithms of Oppression* (New York: NYU Press, 2018); and Cathy O'Neil, *Weapons of Math Destruction* (New York: Random House, 2016).

28. Penny Gillespie and Guneet Bharaj, "Use Personalization to Enrich Customer Experience and Drive Revenue," *Gartner*, April 3, 2019, pp. 6–7.

29. [Staff], "Big Predictions: Agency Gurus on the Future of Data in the Industry," *Media Marketing & Media*, February 1, 2018.

30. Turow, *The Aisles Have Eyes*.

31. Holman W. Jenkins, Jr., "Google and the Search for the Future," *Wall Street Journal*, August 14, 2010, https://www.wsj.com/articles/SB1000 1424052748704901104575423294099527212, accessed October 25, 2018.

32. "Click Fraud," Wikipedia, https://en.wikipedia.org/wiki/Click_fraud, accessed June 3, 2020.

33. "Ad Blocking," Wikipedia, https://en.wikipedia.org/wiki/Ad_blocking, accessed June 4, 2020.

34. George Slefo, "Report: For Every $3 Spent on Digital Ads, Fraud Takes $1," *Advertising Age*, October 22, 2015, http://adage.com/article/digital/ad-fraud-eating-digital-advertising-revenue/301017/, accessed September 13, 2018.

35. Alexandra Bruell, "Fraudulent Web Traffic Continues to Plague Advertisers, Other Businesses," *Wall Street Journal,* March 28, 2018, https://www.wsj.com/articles/fraudulent-web-traffic-continues-to-plague-advertisers-other-businesses-1522234801, accessed February 11, 2018.

36. Juniper Research, "Ad Fraud to Cost Advertisers $19 billion in 2018, Representing 9% of Total Digital Advertising Spend," Business Wire, September 26, 2017, https://www.businesswire.com/news/home/20170926005177/en/Juniper-Research-Ad-Fraud-Cost-Advertisers-19, accessed February 11, 2018.

37. Nicole Perrin, "Demanding a Better Ad Experience: Why One in Four Internet Users Say No to Ads," eMarketer, December 4, 2018, https://content-na1.emarketer.com/demanding-a-better-ad-experience, accessed February 11, 2019.

38. "Gartner Predicts 80% of Marketers Will Abandon Personalization Efforts by 2025," Gartner Press Release, December 2, 2019, https://www.webwire.com/ViewPressRel.asp?aId=250911, accessed November 14, 2020.

39. Susan Moore, "Risk-Averse Privacy Ideas Often Prevent Organizations from Creating Great Customer Experiences," Smarter with Gartner, April 9, 2019, https://www.gartner.com/smarterwithgartner/how-to-balance-personalization-with-data-privacy/, accessed December 5, 2019.

40. Charles Golvin, Benjamin Bloom, and Jennifer Polk, "Predicts 2020: Marketers, They're Just Not That Into You," Gartner Information Technology Research, November 11, 2019, p. 7.

41. Interview with Pete Erickson, June 6, 2019.

42. Tom Shapiro, "How Emotion-Detection Technology Will Change Marketing," HubSpot, October 17, 2016, https://blog.hubspot.com/marketing/emotion-detection-technology-marketing, accessed December 5, 2019.

43. "Clarabridge Unveils New Updates to the Clarabridge Banking Solution," Clarabridge, November 7, 2018, https://www.clarabridge.com/clarabridge-unveils-new-updates-to-the-clarabridge-banking-solution/, accessed December 5, 2019.

44. Angela Chen, "Why Companies Want to Mine the Secrets in Your Voice," The Verge, March 14, 2019, https://www.theverge.com /2019/3/14/18264458/voice-technology-speech-analysis-mental-health-risk-privacy, accessed October 16, 2019.

45. Ibid.

46. Ibid.

47. Ibid.

48. John McCormick, "What AI Can Tell from Listening to You," *Wall Street Journal*, April 1, 2019, https://www.wsj.com/articles/what-ai-can-tell-from-listening-to-you-11554169408, accessed October 16, 2019.

49. Pinelopi Troullinou, "Exploring the Subjective Experience of Everyday Surveillance: The Case of Smartphone Devices as Means of Facilitating 'Seductive' Surveillance," PhD diss., Open University, 2017.

50. Karl Maton, "Habitus," in Michael Grenfell, ed., *Pierre Bourdieu: Key Concepts* (London: Routledge, 2008), pp. 48–64.

51. Tony Bennett and Francis Dodsworth, "Habit and Habituation: Governance and the Social," *Body & Society* 19, nos. 2, 3 (2013).

52. Troullinou, "Exploring the Subjective Experience of Everyday Surveillance," p. 51.

53. Ibid.

54. Andrew McStay, *Emotional AI* (London: Sage 2019), pp. 75, 77.

55. Bret Kinsella, "Why Tech Giants Are so Desperate to Provide Your Voice Assistant," *Harvard Business Review*, March 7, 2019, https://hbr.org/2019/05/why-tech-giants-are-so-desperate-to-provide-your-voice-assistant, October 16, 2019.

56. Quoted in Troullinou, "Exploring the Subjective Experience of Everyday Surveillance," p. 13.

57. Quoted in Nora A. Draper and Joseph Turow, "The Corporate Cultivation of Digital Resignation," *New Media & Society* 2, no. 8 (2019): 1830.

58. Ibid., pp. 1824–1839.

59. Joseph Turow, Michael Hennessy, and Nora Draper, *The Tradeoff Fallacy* (Philadelphia: Annenberg School for Communication, 2015).

60. See Draper and Turow, "Corporate Cultivation of Digital Resignation," p. 3.

1 RISE OF THE SEDUCTIVE ASSISTANTS

1. Emil Protalinski, "Amazon Echo Is a $200 Voice-Activated Wireless Speaker for Your Living Room," VentureBeat, November 6, 2014, https://venturebeat.com/2014/11/06/amazon-echo-is-a-200-voice-activated-smart-wireless-speaker-for-your-living-room/, accessed on March 21, 2019. Excerpted by permission of Emil Protalinski, executive editor of VentureBeat.

2. "Amazon Echo—Now Available," Amazon video pasted into ibid.

3. James O'Toole, "Amazon's Quirky Echo Is Siri in a Speaker," CNN Business, November 6, 2014, https://money.cnn.com/2014/11/06/technology/innovationnation/amazon-echo/index.html, accessed on March 21, 2019.

4. B. H. Juang and Lawrence Rabiner, "Automatic Speech Recognition—A Brief History of the Technology Development," https://www.semanticscholar.org/paper/Automatic-Speech-Recognition-A-Brief-History-of-the-Rabiner/1d199099a2f4f8749c7e10480b29f5adaecad4a1, accessed May 29, 2020.

5. "History of the Cylinder Phonograph," Library of Congress, n.d., https://www.loc.gov/collections/edison-company-motion-pictures-and-sound-recordings/articles-and-essays/history-of-edison-sound-recordings/history-of-the-cylinder-phonograph/. For more on the other early voice devices see Justine Humphry and Chris Chesher, "Preparing for Smart Voice Assistants," *New Media & Society*, May 2020, https://journals.sagepub.com/doi/10.1177/1461444820923679, accessed September 9, 2020.

6. "Speech Recognition," Wikipedia, https://en.wikipedia.org/wiki/Speech_recognition, accessed April 7, 2019.

7. Steve Lohr and John Markoff, "Computers Learn to Listen, and Some Talk Back," *New York Times*, June 24, 2010, https://www.nytimes.com/2010/06/25/science/25voice.html, accessed August 24, 2020.

8. Kjell Carlsson, "Analytical Intelligence Supercharges Speech Analytics," Forrester, September 18, 2017, p. 3.

9. Siri Team, "Deep Learning for Siri's Voice," *Apple Machine Learning Journal* 1, no. 4 (August 2017), https://machinelearning.apple.com/2017/08/06/siri-voices.html, accessed May 29, 2020.

10. Interview with Vlad Sejnoha, May 7, 2019.

11. Lohr and Markoff, "Computers Learn to Listen, and Some Talk Back."

12. Harry Kyriakodis, "Wired City," HiddenCity Philadelphia, January 14, 2013, https://hiddencityphila.org/2013/01/wired-city/#, accessed March 24, 2019.

13. "The History of Call Center Technology," Teledirect, n.d., https://www. teledirect.com/history-call-center-technology-infographic/, accessed March 24, 2019.

14. Nick D'Alleva, "The History of the Call Center Infographic," Specialty Answering Service, April 22, 2015, https://www. specialtyansweringservice.net/the-history-of-the-call-center -infographic/, accessed March 24, 2019.

15. Anna Weinberger of Autonomy, quoted in Paula Bernier, "The History and Advancement of the Contact Center and the Customer Experience," *Customer Interaction Solutions*, August 3, 2012, http:// www.tmcnet.com/cis/features/articles/301771-history-advancement- the-contact-center-the-customer-experience.htm, accessed March 30, 2019.

16. Ibid.

17. "Study: Call Center Issue Resolution Drives Loyalty, Retention, Satisfaction," *Insurance Journal*, July 1, 2008, https://www.insurancejournal. com/news/national/2008/07/01/91519.htm, accessed March 30, 2019.

18. Ibid.

19. Paula Bernier, "Incision Shares 30 Years of Call Center Wisdom," *Customer Interaction Solutions*, August 3, 2012, http://www.tmcnet.com/ cis/departments/articles/301767-infocision-shares-30-years-call- center-wisdom.htm, accessed April 1, 2019.

20. For a discussion of iPBX for contact centers, see "What Is a PBX Phone System?," 3CX, n.d., https://www.3cx.com/pbx/pbx-phone-system/, accessed March 30, 2019.

21. Gene Green and David McKinley, "Why Congress Should Get Behind the Bipartisan 'US Call Center Worker and Protection Act,'" *The Hill*, June 12, 2017, https://thehill.com/blogs/congress-blog/ politics/337368-why-congress-should-get-behind-the-bipartisan-us- call-center, accessed August 24, 2020.

22. "CTI Solutions Inc.," CTI Solutions website, n.d., http://cti-solutions. com/, accessed May 29, 2020.

23. "What Is CTI?," CX Media, n.d., https://cxglobalmedia.com/resources/ faq-items/what-is-cti/, accessed May 29, 2020.

24. Mark E. Andersen, "Call Centers: The Sweatshop of the Modern Era," Daily Kos, January 8, 2012, https://www.dailykos.com/ stories/2012/01/08/1050902/-Call-centers:-The-sweatshop-of-the-modern-era?via=blog_1, accessed April 1, 2019.

25. Ashley Feinberg, "Not Just Comcast: 19 Stories from Call Center Hell," Gizmodo, July 28, 2014, https://gizmodo.com/not-just-comcast-19-stories-from-call-center-hell-1611989865, accessed April 1, 2019.

26. H.R. 1300: United States Call Center Worker and Consumer Protection Act of 2017, https://www.govtrack.us/congress/bills/115/ hr1300, accessed April 1, 2019.

27. Bernier, "Incision Shares 30 Years of Call Center Wisdom."

28. For the Panasonic discussion, see Lohr and Markoff, "Computers Learn to Listen, and Some Talk Back"; the US Airways citation is in Natasha Singer, "The Human Voice, as Game Changer," *New York Times*, April 1, 2012, https://www.nytimes.com/2012/04/01/ technology/nuance-communications-wants-a-world-of-voice-recognition.html, accessed April 8, 2019.

29. Lohr and Markoff, "Computers Learn to Listen, and Some Talk Back."

30. Ibid.

31. One summary of this history is Bianca Bosker, "Siri Rising: The Inside Story of Siri's Origins, and Why She Could Overshadow the iPhone," HuffPost, January 22, 2013, https://www.huffpost.com/entry/siri-do-engine-apple-iphone_n_2499165, accessed April 8, 2019. The details are secret, but it seems that Apple executives quickly regretted the partnership with Nuance and worked for six years to create their own version. See David Pierce, "How Apple Finally Made Siri Sound More Human," *Wired*, September 7, 2017, https://www.wired.com/story/ how-apple-finally-made-siri-sound-more-human/, accessed April 8, 2019.

32. Bosker, "Siri Rising."

33. Austin Carr, "The Inside Story of Jeff Bezos's Fire Phone Debacle," *Fast Company,* January 6, 2015, https://www.fastcompany.com/3039887/under-fire, accessed October 12, 2019.

34. Charles Dugigg, "Is Amazon Unstoppable?," *New Yorker,* October 10, 2019, https://www.newyorker.com/magazine/2019/10/21/is-amazon-unstoppable, accessed October 12, 2019.

35. Ibid.

36. Daisuke Wakabayashi and Nick Wingfield, "Google, Lagging Amazon, Races Across the Threshold into the Home," *New York Times,* October 2, 2019, https://www.nytimes.com/2016/10/03/technology/google-lagging-amazon-races-across-the-threshold-into-the-home.html, accessed October 12, 2019.

37. Richard Yao, "How and Why the Tech Giants Are Fighting for the Home Platform," April 12, 2018, https://medium.com/ipg-media-lab/how-and-why-the-tech-giants-are-fighting-for-the-home-platform-811f5b6174e2, accessed April 15, 2019.

38. Independent analyst Ben Thompson, quoted in James Vincent, "Amazon's Answer to UPS Is Reportedly Getting Its First Proper Launch in LA," The Verge, February 9, 2018, https://www.theverge.com/2018/2/9/16994512/amazon-shipping-service-swa-launching-soon, accessed June 2, 2020.

39. Clifford Nass et al., "Can Computer Personalities Be Human Personalities?," *International Journal of Human-Computer Studies* 43, no. 2 (August 1995): 223–239. See also Andrea Guzman, "Voices in and of the Machine," *Computers in Human Behavior* 90 (2019): 343–350.

40. Phi Tran, "Cell Phone Inventor Martin Cooper Was Influenced by Star Trek," *Adweek,* April 3, 2013, https://www.adweek.com/digital/cell-phone-maker-martin-cooper-was-influenced-by-star-trek/, accessed June 3, 2020.

41. Judith Shulevitz, "Alexa, Should We Trust You?," *Atlantic,* November 2018, https://www.theatlantic.com/magazine/archive/2018/11/alexa-how-will-you-change-us/570844/, accessed April 9, 2019.

42. Surur, "Petition for Microsoft to Make [the] Code Name Cortana Official Has 2000 votes," MSPowerUser, September 13, 2013, https://

mspoweruser.com/petition-for-microsoft-to-make-to-code-name-cortana-official-already-has-2000-votes/, accessed April 10, 2019.

43. Tom Warren, "The Story of Cortana, Microsoft's Siri Killer," The Verge, April 2, 2014, https://www.theverge.com/2014/4/2/5570866/cortana-windows-phone-8–1-digital-assistant; and ibid.

44. Adam Clark Estes, "Uh, Apple, Did You Think This Through?," June 5, 2017, https://gizmodo.com/uh-apple-did-you-think-this-through-1795830661; "Apple HomePod Review," T3, n.d., https://www.t3.com/reviews/apple-homepod-review; and "HomePod Sounds Great—But Is There Any Way to Shut Siri Up?," Reddit, https://www.reddit.com/r/apple/comments/81acwy/homepod_sounds_greatbut_is_there_any_way_to_shut/, accessed April 10, 2019.

45. Jon Fingas, "Google Plots a Backstory for Its AI Assistant," engadget, https://www.engadget.com/2016-05-31-google-assistant-personality.html, May 30, 2020.

46. Matthew Lynley, "Google Unveils Google Assistant, a Virtual Assistant That's a Big Upgrade to Google Now," TechCrunch, May 18, 2016, https://techcrunch.com/2016/05/18/google-unveils-google-assistant-a-big-upgrade-to-google-now/, accessed May 30, 2020.

47. Rowland Manthorpe, "The Human (and Pixar Characters) Inside Google's Assistant," Wired, February 2017, https://www.wired.co.uk/article/the-human-in-google-assistant, accessed April 12, 2019.

48. Ben Fox Rubin," Alexa, Be More Human: Inside Amazon's Effort to Make Its Voice Assistant Smarter, Chattier, and More Like You," CNET, August 29, 2017, https://www.cnet.com/special-reports/amazon-alexa-echo-inside-look/, April 14, 2019.

49. Bianca Bosker, "Why Siri's Voice Is Now a Man (and a Woman)," HuffPost, June 12, 2013, https://www.huffpost.com/entry/siri-voice-man-woman_n_3423245, accessed April 13, 2019.

50. Chandra Steele, "The Real Reason Voice Assistants Are Female (and Why It Matters)," PC Magazine, January 4, 2018, https://www.pcmag.com/commentary/358057/the-real-reason-voice-assistants-are-female-and-why-it-matt, accessed April 13, 2019.

51. Alejandro Alba, "Check Out How Samsung Describes Its Male and Female Bixby Assistants [Updated]," Gizmodo, July 19, 2017, https://

gizmodo.com/check-out-how-samsung-describes-its-male-and-female-bix-1797051987, accessed April 13, 2019.

52. Quoted in Joanna Stern, "Alexa, Siri, Cortana: The Problem with All-Female Digital Assistants," *Wall Street Journal,* February 21, 2017, https://www.wsj.com/articles/alexa-siri-cortana-the-problem-with-all-female-digital-assistants-1487709068, accessed July 15, 2020.

53. "Hands-Free Help from Google Home," YouTube, n.d., https://www.youtube.com/watch?v=-C2BQUhn3IQ, accessed April 13, 2019.

54. Google, "Google Privacy Policy: Information We Collect as You Use Our Services," Google, July 1, 2020, https://policies.google.com/privacy. See also Google, "Data Security and Privacy on Devices that Work with Assistant," n.d., https://support.google.com/googlenest/answer/7072285?hl=en, accessed, July 15, 2020.

55. Nick Staff, "Why Google's Fancy New Assistant Is Just Called 'Google,'" The Verge, May 20, 2016, https://www.theverge.com/2016/5/20/11721278/google-ai-assistant-name-vs-alexa-siri, accessed April 15, 2019.

56. Brett Kinsella, "Google Duplex Shows the Way to a True Virtual Assistant," Voicebot.ai, May 8, 2018, https://voicebot.ai/2018/05/08/google-duplex-shows-the-way-to-a-true-virtual-assistant/, accessed April 13, 2019.

57. Ron Amadeo, "Talking to Google Duplex: Google's Human-like Phone AI Feels Revolutionary," Ars Technica, June 27, 2018, https://arstechnica.com/gadgets/2018/06/google-duplex-is-calling-we-talk-to-the-revolutionary-but-limited-phone-ai/3/, accessed April 15, 2019. See also Kyle Wiggers, "Google Duplex's Haircut-Booking Feature Could Help Short-Handed Businesses," VentureBeat, October 12, 2020, https://venturebeat.com/2020/10/12/google-duplexs-haircut-booking-feature-could-help-short-handed-businesses/, accessed November 15, 2020.

58. Leslie Joseph, "The Six Factors That Separate Hype from Hope in Your Conversational AI Journey," Forrester Research, August 6, 2018, https://www.forrester.com/report/The+Six+Factors+That+Separate+Hype+From+Hope+In+Your+Conversational+AI+Journey/-/E-RES143773, accessed August 24, 2020.

59. Vlad Sejnoha, "Beyond Voice Intelligence: It's the Age of Intelligent Systems," *Forbes*, January 11, 2013, https://www.forbes.com/sites/ciocentral/2013/01/11/beyond-voice-recognition-its-the-age-of-intelligent-systems/#5ce1e88d1ad3, accessed April 28, 2019.

60. See "Contact Center AI," Google, n.d., https://cloud.google.com/solutions/contact-center/; Emily Winchurch, "Announcing the AI Solution with Voice for Contact Centers," Watson Blog, IBM, September 21, 2018, https://www.ibm.com/blogs/watson/2018/09/announcing-the-ai-solution-with-voice-for-contact-centers/, and Rob Koplowitz et al., "The Forrester New Wave™: Conversational Computing Platforms, Q2 2018," Forrester Research, April 12, 2018, https://reprints.forrester.com/#/assets/2/73/RES137816/reports, all accessed April 21, 2019.

61. "Customer Care Voice Agent," IBM Watson, n.d., https://www.ibm.com/watson/assets/duo/pdf/Customer_Care_Voice_Agent.pdf, accessed April 21, 2019.

62. "Intelligent Automated Conversation with a Human Touch," Nuance, https://www.nuance.com/omni-channel-customer-engagement/digital/virtual-assistant/nina.html#, accessed May 1, 2019.

63. "Contact Center AI," Google, n.d., https://cloud.google.com/solutions/contact-center/; Winchurch, "Announcing the AI Solution with Voice for Contact Centers."

64. "Forrester Q&A: Customer Journey Analytics," Nuestar, n.d., https://www.home.neustar/resources/faqs/customer-journey-analytics-forrester-q-a, accessed December 15, 2019.

2 WHAT MARKETERS SEE IN VOICE

1. Ahmed Fuad Siddiqui (Inventor) for Amazon Technologies, Inc, "User Identification Using Voice Characteristics," US Patent 10,262,661, April 16, 2019.

2. David Gustafson, "Additional U.S. Patent Highlights Mattersight's Predictive Behavioral Routing Capabilities," Mattersight, May 7, 2014, https://www.globenewswire.com/news-release/2014/05/07/919167/0/en/Additional-U-S-Patent-Highlights-Mattersight-s-Unique-Predictive-Behavioral-Routing-Capabilities.html, accessed September 13, 2020.

3. "Benefits," Nuance, https://www.nuance.com/omni-channel-customer-engagement/security/identification-and-verification.html, accessed May 9, 2019.

4. See "Alexa, Echo Devices, and Your Privacy," Amazon, n.d., https://www.amazon.com/gp/help/customer/display.html?nodeId=GVP69FUJ48X9DK8V; and Jason Cipriani, "Amazon Echo Stores Your Voice Commands. Here's How Alexa Can Delete Them," CNET, May 29, 2020, https://www.cnet.com/how-to/amazon-echo-stores-your-voice-commands-heres-how-alexa-can-delete-them/, accessed July 15, 2020.

5. "Other Places Voice and Audio Recordings May Be Saved," n.d., https://support.google.com/accounts/answer/6030020?co=GENIE.Platform%3DDesktop&hl=en, accessed July 15, 2020.

6. Interview with Joe Petro, July 19, 2019.

7. Rupert Jones, "Voice Recognition: Is It Really as Secure as It Sounds?," *Guardian*, September 22, 2018, https://www.theguardian.com/money/2018/sep/22/voice-recognition-is-it-really-as-secure-as-it-sounds, accessed May 10, 2019.

8. Wayne Rash, "Nuance Biometric Security Turns Your Body into an Authentication Tool," eWeek, February 28,2018, https://www.eweek.com/security/nuance-biometric-security-turns-your-body-into-an-authentication-tool; and "Making Voice Biometrics Harder to Hack," August 29, 2018, Pymnts.com, https://www.pymnts.com/news/security-and-risk/2018/hackers-biometrics-cybercrime-fraudsters-authentication/.

9. Jones, "Voice Recognition."

10. Discussion with Rita Singh, Carnegie Mellon University, March 28, 2019.

11. Rita Singh, *Profiling Humans from Their Voice* (New York: Springer, 2019), p. vii.

12. Ibid., p. 86.

13. Ibid., p. 110.

14. Ibid., p. 88.

15. Ibid., p. 89.

16. Ibid., p. 98.

17. Ibid., p. 102.
18. Ibid.
19. Ibid., p. 103.
20. Ibid., p. 104.
21. Ibid., pp. 117–118.
22. Mindy Weisberger, "AI Listened to People's Voices. Then It Generated Their Faces," LiveScience, June 11, 2019, https://www.livescience.com/65689-ai-human-voice-face.html, accessed October 23, 2019.
23. Singh, *Profiling Humans*, p. 109.
24. For example, research on noting a person's social status from "brief speech" took place at Yale University's School of Management. See Michael W. Kraus, Brittany Torrez, Jun Won Park, and Fariba Ghayebi, "Evidence for the Reproduction of Social Class in Brief Speech, *PNAS*, September 22, 2019, https://www.pnas.org/content/early/2019/10/15/1900500116, accessed October 22, 2019.
25. Rita Singh, personal communication, April 22, 2019.
26. Singh, *Profiling Humans*, p. 3.
27. Interview with Rita Singh, March 28, 2019.
28. See, for example, James D. Harnsberger et al., "Stress and Deception in Speech: Evaluating Layered Voice Analysis," *Journal of Forensic Science* 54, no. 3 (May 2009): 642–639; and Darren Haddad et al., "Investigation and Evaluation of Voice Stress Analysis Technology," Final Report to the Office of Justice Programs, U.S. Department of Justice, February 13, 2002, https://www.ncjrs.gov/pdffiles1/nij/193832.pdf, accessed May 14, 2019.
29. Singh, *Profiling Humans*, p. 109.
30. George I. Seffers, "The Mind-Blowing Promise of AI-Driven Voice Profiling," *Signal*, September 1, 2018, https://www.afcea.org/content/mind-blowing-promise-ai-driven-voice-profiling, accessed May 30, 2019. The company's thin website in mid-2019 was more ambitious, claiming that "no matter the application—forensics, medical diagnosis, fraud detection, security, and more—we can help. Send us a note . . . and we'll be in touch."
31. Heather Mack, "Beyond Verbal Launches API to enable Voice-Based Emotion Detection by Virtual Private Assistants," Mobile Health News,

https://www.mobihealthnews.com/content/beyond-verbal-launches-api-enable-voice-based-emotion-detection-virtual-private-assistants, accessed September 6, 2020.

32. "VocalisHealth," https://vocalishealth.com/, accessed July 19, 2020.

33. Ibid.

34. Elad Maor et al., "Voice Signal Characteristics Are Independently Associated with Coronary Artery Disease," *Mayo Clinic Proceedings* 93, no. 7 (April 2018): 840–847, quotation on p. 846.

35. "VocalisHealth," https://vocalishealth.com/, accessed July 18, 2020.

36. "Emotions Win," Invoca and Adobe, n.d., https://www.invoca.com/brand-eq/#start, p. 20, accessed May 29, 2019.

37. Ian Jacobs and Kjell Carlsson, "New Tech: AI-Fueled Speech Analytics Solutions, Q2, 2018," Forrester Research, June 21, 2016, p. 2, https://www.forrester.com/report/New+Tech+AIFueled+Speech+Analytics+Solutions+Q2+2018/-/E-RES142152, accessed August 24, 2020.

38. "DialogTech and Amazon Web Services Enable Marketers and Contact Centers to Grow Revenue," DialogTech, n.d., https://www.dialogtech.com/blog/dialogtech-amazon-web-services/, accessed October 22, 2019.

39. Kjell Carlsson, "Artificial Intelligence Supercharges Speech Analytics," Forrester, September 18, 2017, p. 2, https://www.forrester.com/report/Artificial+Intelligence+Supercharges+Speech+Analytics/-/E-RES139795, accessed August 24, 2020.

40. "How AI Improves the Customer Experience," CallMiner Eureka, n.d., https://www.technoreports.info/resourcefiles/HOW_AI_IMPROVES_THE_CUSTOMER_EXPERIENCE.pdf, accessed August 24, 2020.

41. "Nexidia Analytics," n.d., NICE website, https://www.nice.com/engage/nexidia-customer-engagement-analytics/nexidia-interaction-analytics, accessed October 22, 2019.

42. "Analyze Conversations for Actionable Marketing Insights," Dialogtech, n.d., https://www.dialogtech.com/products/analytics, accessed May 30, 2019.

43. "How It Works," CallMiner, n.d., https://callminer.com/, accessed on May 17, 2019.

44. "Enhanced Voice Customer Analytics with CallMiner," n.d., https://learn.callminer.com/wistia-product-videos/voice-of-customer-speech-analytics, accessed October 22, 2019.

45. "How It Works," CallMiner, n.d., https://callminer.com/products/analyze/#, accessed October 22, 2019.

46. "Speech Analytics Transcription Accuracy," Verint, [2018], https://www.verint.com/wp-content/uploads/speech-analytics-transcription-accuracy-white-paper-english-us.pdf, accessed May 18, 2019.

47. "Customer Experience," CallMiner, n.d., https://callminer.com/solutions/business-value/customer-experience/, accessed June 5, 2020.

48. Discussion with Scott Eller, Founder, Neuraswitch, May 15, 2019.

49. "Contact Center Innovations Real-Time Speech Analytics & Knowledge Management in Call Centers," Verint, n.d., http://video.verint.com/watch/WhUB8fr5RD2NJoSaABAMci, accessed May 21, 2019.

50. Cogito Corporation, "Cogito for the Call Center," n.d., https://www.cogitocorp.com/, accessed May 18, 2019.

51. Allison Snow et al., "B2B Insights Deep Dive," Forrester, August 8, 2019, p. 5, https://www.forrester.com/report/B2B+Insights+Deep+Dive+Use+AI+To+Drive+Engagement/-/E-RES149717, accessed August 24, 2020.

52. "Drive Better Phone Interactions with Artificial Intelligence," Cogito, n.d., https://www.cogitocorp.com/product/, accessed May 28, 2019.

53. Greg Nichols, "Artificial Empathy: Call Center Employees Are Using Voice Analytics to Predict How You Feel," ZDNet, March 20, 2018, https://www.zdnet.com/article/artificial-empathy-call-center-employees-are-using-voice-analytics-to-predict-how-you-feel/, accessed May 28, 2019.

54. Interview with Andy Traba, May 8, 2019.

55. Ibid.

56. Ibid.

57. "Neuraswitch and Beyond Verbal Announce Partnership to Add Emotion AI Insights to Call-Center Customer Experience Platform," Beyond Verbal Communications, October 31, 2018, http://www.digitaljournal.com/pr/4004199?noredir=1, accessed August 24, 2020.

58. Traba interview.

59. "Google Privacy Policy," Google, March 31, 2020, https://policies. google.com/privacy?hl=en-US, accessed May 31, 2020.

60. "Our Commitment to Privacy in the Home," Google, n.d., https://store. google.com/us/magazine/google_nest_privacy, accessed May 31, 2020.

61. "Alexa and Alexa Device FAQs," Amazon, n.d., https://www.amazon. com/gp/help/customer/display.html?nodeId=201602230, accessed March 20, 2020.

62. Alexa Terms of Use, January 1, 2020, https://www.amazon.com/gp/ help/customer/display.html?nodeId=201602230, accessed March 20, 2020.

63. Anonymous interview, March 31, 2019.

64. Interview with Jeremy Carney, May 20, 2019.

65. Interview with Tom Cowan, May 6, 2019.

66. Jim Huafeng and Sho Wang, "Voice-Based Determination of Physical and Emotional Characteristics of Users," Amazon Technologies Inc., US Patent US10096319B1, 2017, https://patents.google.com/patent/ US10096319B1/en, accessed June 5, 2020.

67. Anthony Fadell et al. for Google Inc., "Smart Home Automation System That Suggests or Automatically Implements Selected Household Policies Based on Sensed Observations," US patent 10,114,351, October 30, 2018, https://patents.google.com/patent/ US10114351B2/en, accessed August 24, 2020.

68. Ibid.

69. Brian Lafayette Adair for Amazon Technologies, Inc., "Indirect Feedback Systems and Methods," US patent 10,019,489, July 10, 2018, column 2, https://patents.google.com/patent/US10019489B1/ en, accessed August 24, 2020.

70. Matthew Sharifi and Jakob Foerster for Google LLC, "Adaptive Text-to-Speech Outputs," US Patent 10,453,441, October 22, 2019, https:// patents.google.com/patent/US10453441B2/en, accessed July 15, 2020.

71. Anthony Michael Fadell et al., "Visitor Options at an Entryway to a Smart-Home," US Patent 9,978,238, May 22, 2018, https://patents. google.com/patent/US9978238B2/en, accessed August 24, 2020.

72. Ying Zheng for Google LLC, "Systems and Methods for Associating Media Content with Viewer Expressions," US Patent 9,866,903, January 9, 2018, http://patft.uspto.gov/netacgi/nph-Parser?Sect1=PT

O1&Sect2=HITOFF&d=PALL&p=1&u=%2Fnetahtml%2FPTO%2Fsrc
hnum.htm&r=1&f=G&l=50&s1=9,866,903.PN.&OS=PN/9,866,903&R
S=PN/9,866,903, accessed September 6, 2020.

73. Eric Meissner et al. for Amazon, Inc., "Contextual Presence," US
Patent 10,455,362, October 22, 2019, https://patents.justia.com/
patent/10455362, accessed August 24, 2020.

74. For example, Bret Kinsella, "RBC Analyst Says 52 Million Google
Home Devices Sold to Date and Generating $3.4 Billion in 2018
Revenue," Voicebot.ai, December 24, 2018, https://voicebot.
ai/2018/12/24/rbc-analyst-says-52-million-google-home-devices-sold-
to-date-and-generating-3-4-billion-in-2018-revenue/, accessed July
15, 2020; and Greg Sterling, "Google Takes Baby Steps to Monetize
Google Assistant, Google Home," Search Engine Land, April 22, 2019,
https://searchengineland.com/google-takes-baby-steps-to-monetize-
google-assistant-google-home-315743, accessed July 15, 2020.

75. Eugene Kim, "As Amazon Floods the Market with Alexa Devices, the
Business Model Is Getting Fresh Scrutiny," CNBC.com, September 28,
2019, https://www.cnbc.com/2019/09/28/amazon-alexa-growth-has-
investors-questioning-the-business-model.html, accessed July 25, 2020.

76. Ibid.

77. Interview with Brandon Purcell, June 21, 2019.

3 AN OPERATING SYSTEM FOR YOUR LIFE

1. Nick Statt, "Amazon Wants Alexa to Be the Operating System for
Your Life," The Verge, September 27, 2018, https://www.theverge.
com/2018/9/27/17911300/amazon-alexa-echo-smart-home-eco-
system-competition, accessed August 4, 2019.

2. Chris Welsh, "The 14 Biggest Announcements from Amazon's
Surprise Hardware Event," The Verge, September 20, 2018, https://
www.theverge.com/2018/9/20/17883242/amazon-alexa-event-2018-
news-recap-echo-auto-dot-sub-link-auto-microwave, accessed August
4, 2019.

3. "Amazon Echo Integration with Ring Devices," Ring, n.d., https://
support.ring.com/hc/en-us/articles/115003247146-Integrating-
Amazon-Alexa-supported-Devices-with-Ring-Devices, accessed
September 13, 2020.

4. Statt, "Amazon Wants Alexa to Be the Operating System for Your Life."

5. Jennifer Wise, quoted in ibid.

6. Tony Bennett and Francis Dodsworth, "Habit and Habituation: Governance and the Social," *Body & Society* 19, nos. 2, 3 (2013).

7. Jennifer Wise, quoted in Statt, "Amazon Wants Alexa to be the Operating System for Your Life."

8. Andria Cheng, "Why This Year's Prime Day Is Even More Crucial for Amazon," *Forbes,* July 16, 2018, https://www.forbes.com/sites /andriacheng/2018/07/16/this-is-the-single-most-important-watchpoint- for-amazons-prime-day/#717ab1d863a1, accessed August 18, 2020.

9. "Over 17 Million Transactions Made on Amazon over Prime Day 2018," Hitwise, July 20, 2018, https://www.hitwise.com/ en/2018/07/20/over-17-million-transactions-made-on-amazon-over- prime-day-2018/, accessed August 5, 2019.

10. Kristin McGrath, "What Is Prime Day?," BlackFriday.com, July 18, 2019, https://blackfriday.com/news/amazon-prime-day-history-and- statistics, accessed August 5, 2019.

11. Ibid.

12. Ibid.

13. "Alexa, How Was Prime Day? Prime Day 2019 Surpassed Black Friday and Cyber Monday Combined," Amazon Press Release, July 17, 2019, https://press.aboutamazon.com/news-releases/news-release-details/ alexa-how-was-prime-day-prime-day-2019-surpassed-black-friday, accessed August 5, 2019.

14. Drake Hawkins, "Pre-Prime Day Amazon Deal on Ecobee4 Smart Thermostat," DigitalTrends, July 3, 2019, https://www.digitaltrends. com/dtdeals/ecobee4-smart-thermostat-pre-amazon-prime-day-deal/, accessed August 24, 2020.

15. Interview with Brad Russell, July 25, 2019.

16. The information from NMG in this and the following paragraphs are from my interview with Tom Hickman, July 25, 2019.

17. Kyle Wiggers, "Adam Cheyer: Samsung's Plan for Winning with Bixby Is Empowering Third-Party Developers," VentureBeat, June 22, 2019, https://venturebeat.com/2019/06/22/adam-cheyer-samsungs-plan- for-winning-with-bixby-is-empowering-third-party-developers/, accessed August 24, 2020. Cheyer left the project in 2020.

18. "Control4 Smart Home OS3," Control4, https://www.control4.com/os3/, accessed July 30, 2019.

19. "End User License Agreement," Control4, https://www.control4.com/legal/end-user-license-agreement, accessed March 22, 2010.

20. Toll Brothers, "Toll Brothers Expands Smart Home Technology Available in Homes," Global NewsWire, June 26, 2018, https://www.globenewswire.com/news-release/2018/06/26/1529745/0/en/Toll-Brothers-Expands-Smart-Home-Technology-Available-in-Homes.html, accessed August 8, 2019.

21. "Be Smart. Live Easier. KB Smart Home System with Google," KB Homes, n.d., https://www.kbhome.com/kb-smart-home, accessed August 8, 2019.

22. "Wake Up the Home!" video, Lennar, n.d., https://www.lennar.com/ei/connectivity, accessed August 8, 2019.

23. "Lennar Introduces World's First CERTIFIED™ Home Designs," https://www.lennar.com/wifi-certified, accessed September 7, 2020.

24. "Amazon Smart Home Activation: Lennar," n.d., https://www.amazon.com/gp/product/B075311V17, accessed August 8, 2019.

25. Quotations from Tony Callahan in this and the following paragraphs are from my interview with him on July 17, 2019.

26. The Shea findings parallel broader research by the Voicebot.ai news and consulting firm. See Bret Kinsella, "Voice Industry Professionals Say Amazon Alexa Is Having the Biggest Impact Followed by Google with Everyone Else Far Behind—New Report," Voicebot.ai, May 11, 2020, https://voicebot.ai/2020/05/11/voice-industry-professionals-say-amazon-alexa-is-having-the-biggest-impact-followed-by-google-with-everyone-else-far-behind-new-report/, accessed May 11, 2020.

27. Sasha Lekach, "BMW Makes Sure We Can't Escape Voice Assistants While Driving," *Mashable,* September 7, 2018, https://mashable.com/article/bmw-digital-assistant-car-voice/, accessed August 14, 2019.

28. Interview with Eric Montague, July 18, 2019.

29. Robert Bruchhardt, "Three Things We Learned Building the Hey Mercedes Voice Assistant," a talk presented at the Voice 2019 conference, July 24, 2019. See https://www.voicesummit.ai/agenda, accessed August 11, 2019.

30. Patrick Gaelweiler, "What's That Place? Open That Window! Play That Song," Nuance press release, January 23, 2019, https://www.cerence.com/news-releases/news-release-details/whats-place-open-window-play-song-experiencing-digital-car, accessed September 7, 2020.

31. Bret Kinsella, "3-in-5 Consumers Want the Same Voice Assistant in the Car as in the Home—New Amazon and JD Power Study," Voicebot.ai, April 9, 2019, https://voicebot.ai/2019/04/09/3-in-5-consumers-want-the-same-voice-assistant-in-the-car-as-in-the-home-new-amazon-and-jd-power-study/, accessed November 20, 2019.

32. Ibid.

33. Sasha Lekach, "Amazon's Alexa Will Soon Do Your Bidding Through Your Car Infotainment Console," *Mashable*, August 9, 2018, https://mashable.com/article/amazon-alexa-car-voice-integration/, accessed August 14, 2019.

34. Bret Kinsella, "GM to Provide the First Full Alexa Auto Implementation and It's Different Than What Came Before," Voicebot.ai, September 29, 2019, https://voicebot.ai/2019/09/29/gm-to-provide-the-first-full-alexa-auto-implementation-and-its-different-than-what-came-before/, accessed November 20, 2019.

35. Bret Kinsella, "Amazon, Baidu, Cerence, Microsoft, Tencent, and 30 Other Companies Launch Voice Interoperability Initiative," September 24, 2019, https://voicebot.ai/2019/09/24/amazon-baidu-cerence-microsoft-tencent-and-30-other-companies-launch-voice-interoperability-initiative/, accessed November 20, 2019.

36. Quotations from Joe Petro in this and the following paragraphs are from my interview with him on July 19, 2019.

37. Chris Burt, "Auto Biometrics Market Projected for Massive Growth as Industry Players Demo New Technology," January 15, 2020, https://www.biometricupdate.com/202001/auto-biometrics-market-projected-for-massive-growth-as-industry-players-demo-new-technology, accessed January 15, 2020.

38. Kevin Lisota, "Amazon Signals Big Ambitions for Automobiles with Expanded Presence at CES," GeekWire, January 10, 2020, https://www.geekwire.com/2020/amazon-signals-big-ambitions-automobiles-expanded-presence-ces/, accessed March 25, 2020.

39. Jaclyn Trop, "The Spy Inside Your Car," *Fortune,* January 24, 2019, https://fortune.com/2019/01/24/the-spy-inside-your-car/, accessed December 13, 2019.

40. Urvaksh Karkaria, "BMW Introduces Intelligent Personal Assistant Communication System," *Automotive News Europe,* September 6, 2018, https://europe.autonews.com/article/20180906/COPY/309069934/bmw-introduces-intelligent-personal-assistant-communications-system, accessed December 13, 2019.

41. Trop, "Spy Inside Your Car."

42. Ibid.

43. "Information Notice—Speech Messaging," Volvo, updated November 24, 2019, https://www.volvocars.com/mt/support/topics/legal-documents/privacy/privacy-notice---speech-messaging, accessed July 16, 2020.

44. Interview with Roger Lanctot, August 12, 2019.

45. Tamra Johnson, "Think You're in Your Car More? You're Right. Americans Spend 70 Billion Hours Behind the Wheel," AAA Newsroom, February 27, 2019, https://newsroom.aaa.com/2019/02/think-youre-in-your-car-more-youre-right-americans-spend-70-billion-hours-behind-the-wheel/, accessed July 16, 2020.

46. Lanctot interview.

47. William Boston and Tim Higgins, "The Battle for the Last Unconquered Screen—The One in Your Car," *Wall Street Journal,* April 8, 2019, https://www.wsj.com/articles/the-battle-for-the-last-unconquered-screenthe-one-in-your-car-11554523220, accessed June 6, 2020.

48. Ibid.

49. Two anonymous interviews, July 2019.

50. Boston and Higgins, "Battle for the Last Unconquered Screen."

51. Ibid.

52. Anonymous interview, July 21, 2019.

53. Tom Franklin, "Hands-Free Hotel Stays, with Google," Google: The Keyword, August 26, 2020, https://www.blog.google/products/assistant/hands-free-hotel-stays-google/, accessed August 29, 2020.

54. "Alexa for Hospitality," Amazon Alexa, n.d., https://www.amazon.com/alexahospitality, accessed July 16, 2020.

55. Ibid., accessed August 16, 2019.

56. "What to Know About Amazon Alexa in Hotels," Employers, n.d., https://www.employers.com/resources/blog/2019/what-to-know-about-amazon-alexa-in-hotels, accessed November 17, 2019.

57. Christina Jelski, "Amazon's Alexa Can Be an Unwelcome Roommate," *Travel Weekly*, February 19, 2019, https://www.travelweekly.com/Travel-News/Hotel-News/Hotel-guests-uncomfortable-with-Amazon-Alexa, accessed November 19, 2019.

58. Lorraine Sileo, senior vice president, research and business operations at Phocuswright, in ibid.

59. "Alexa for Hospitality."

60. Franklin, "Hands-Free Hotel Stays, with Google."

61. "Angie's Features and Capabilities," Angie Hospitality, n.d., https://angie.ai, accessed December 12, 2019.

62. Jesse Tenne Fox, "Google, Volara Partner for Contactless Hotel Tech," Hotel Management, August 27, 2020, https://www.hotelmanagement.net/tech/google-volara-partner-for-contactless-hotel-tech, accessed September 8, 2020.

63. David Berger, "Lessons Learned from a Failed Amazon Alexa Deployment," *FocusWire*, April 10, 2019, https://www.phocuswire.com/Berger-Volara-on-Alexa-mistakes, accessed July 16, 2020.

64. Interview with Dave Berger, August 16, 2019.

65. Alexa for Hospitality Q&A, n.d., https://m.media-amazon.com/images/G/01/asp-marketing/Alexa_for_Hospitality_FAQs._CB1529382359_.pdf, accessed August 16, 2019.

66. Interview with Aparna Ramanathan, August 16, 2019.

67. See also the interview "Alexa in Education with Dr. Aparna and Deepak Ramathan," October 16, 2018, https://alexaincanada.ca/alexa-in-education-with-aparna-and-deepak-ramanathan/, accessed March 25, 2020.

68. "Ask My Class," https://www.goaskmyclass.com, accessed August 16, 2019. A later iteration (accessed on March 25, 2020) is https://askmyclass.app/.

69. Mark Lieberman, "Using Amazon Echo, Google Home to Learn: Skill of the Future or Bad Idea?," *Education Week*, February 4, 2020, https://www.edweek.org/ew/articles/2020/02/05/using-amazon-echo-google-home-to-learn.html, accessed February 10, 2020.

70. Ibid.

71. Ramanathan interview.

72. Lieberman, "Using Amazon Echo, Google Home to Learn."

73. Ibid.

74. Ibid.

75. "Examples of Information Collected," Amazon Privacy Notice, January 1, 2020, https://www.amazon.com/gp/help/customer/display.html/ref=hp_bc_nav?ie=UTF8&nodeId=201909010#GUID-1B2BDAD4–7ACF-4D7A-8608-CBA6EA897FD3__SECTION_87C837F9CCD84769B4AE2BEB14AF4F01, accessed July 16, 2020.

76. "What About Advertising?," Amazon Privacy Notice, January 1, 2020, https://www.amazon.com/gp/help/customer/display.html/ref=hp_bc_nav?ie=UTF8&nodeId=201909010#GUID-1B2BDAD4–7ACF-4D7A-8608-CBA6EA897FD3__SECTION_87C837F9CCD84769B4AE2BEB14AF4F01, accessed July 16, 2020.

77. Patrick Givens presentation at the Voice 2019 conference, Newark, NJ, July 23, 2019.

78. "Need a Way to Better Assist Shoppers in Store?," SmartAisle, n.d., https://www.smartaisle.io/, accessed August 16, 2019.

79. Mars Agency, "Bottle Genius Demo," SmartAisle, n.d., https://www.smartaisle.io/, accessed August 16, 2019.

80. The Mars Agency (press release), "The Mars Agency and Bottlerocket Wine & Spirit Launch Groundbreaking Voice Activated In-Store Shopper Tool," Cision PR Newswire, February 1, 2018, https://www.prnewswire.com/news-releases/the-mars-agency-and-bottlerocket-wine--spirit-launch-groundbreaking-voice-activated-in-store-shopper-tool-300591675.html, accessed August 24, 2020.

81. Interview with Peter Peng, August 24, 2019.

82. Interview with Ethan Goodman, August 21, 2019.

83. Discussion with Bret Kinsella, April 19, 2019.

4 VOICE TECH CONQUERS THE PRESS

1. "Amazon and Lennar Team Up to Show and Sell Smart Home Tech," CNBC, May 22, 2018, https://www.cnbc.com/video/2018/05/23/amazon-lennar-homes-alexa-real-estate-tech.html, accessed May 23, 2019.

2. Judith Shulevitz, "Alexa, Should We Trust You?," *Atlantic,* November 2018, https://www.theatlantic.com/magazine/archive/2018/11/alexa-how-will-you-change-us/570844/, accessed November 30, 2019.

3. Penelopi Troullinou, "Exploring the Subjective Experience of Everyday Surveillance: The Case of Smartphone Devices as Means of Facilitating 'Seductive' Surveillance," PhD diss., Open University, 2017, p. 48.

4. Jenna Wortham, "Technology Innovator's Mobile Move," June 27, 2012, https://www.nytimes.com/2010/06/28/technology/28sri.html, accessed November 30, 2019.

5. Steve Lohr, "Speech Recognition's Early Days," *New York Times,* June 25, 2010, https://www.nytimes.com/2010/06/25/science/25history.html, accessed November 30, 2019.

6. Wortham, "Technology Innovator's Mobile Move."

7. Claire Cain Miller, "BITS: Better Voice Commands for Android Phones," *New York Times,* August 16, 2010, https://archive.nytimes.com/query.nytimes.com/gst/fullpage-9D05E1DB-1338F935A2575BC0A9669D8B63.html, accessed November 30, 2019.

8. Steve Lohr and John Markoff, "Computers Learn to Listen, and Some Talk Back," *New York Times,* June 24, 2010, https://www.nytimes.com/2010/06/25/science/25voice.html, accessed November 22, 2019.

9. Ibid.

10. Timothy Hay, "Speech Recognition Starts to Make Noise," *Wall Street Journal,* April 21, 2010, p. B5A.

11. Jeff Gelles, "Apple Rolls Out New iPhone, Software," *Philadelphia Inquirer,* October 5, 2011, p. A13.

12. Jefferson Graham, "Want an iPhone 4S? Get in Line," *USA Today,* October 14, 2011, p. B1.

13. Jefferson Graham, "Siri's Becoming Everybody's Buddy," *USA Today,* October 21, 2011, p. B1.

14. "Siri, Iris, and the Dream of Just Talking to Our Phones," CNN Business, October 26, 2011.

15. Jefferson Graham, "Some Consumers Love Siri, Others Not so Much," *USA Today,* February 22, 2012, p. B3.

16. Daisuke Wakabayashi and Alistair Barr, "Apple and Google Know What You Want Before You Do," *Wall Street Journal,* August 3, 2015.

17. "In Toronto, Surprises, Stars and Standing O's," *USA Today*, September 9, 2013, p. D4.

18. Sarah LaTrent, "'Her' and the Realities of Computer Love," CNN Wire, January 7, 2014.

19. Melena Ryzik, "Asking Siri About 'Her,'" *New York Times* Blogs, January 26, 2014.

20. Alex Hawgood, "'Interactive' Gets a New Meaning," *New York Times*, December 26, 2013, accessed November 30, 2019, https://www.nytimes.com/2013/12/26/fashion/Sex-toys-cybersex-high-tech.html, accessed November 30, 2019.

21. Robert Kawakami, "How Real Is Spike Jonze's 'Her'?," *Wall Street Journal*, January 24, 2014.

22. Steve Rosenbush, "Google DeepMind Hastens Computers That Think Like People," *Wall Street Journal* CIO Journal, January 31, 2014, https://blogs.wsj.com/cio/2014/01/31/google-deepmind-deal-hastens-computers-that-think-like-people/, accessed August 31, 2019.

23. Richard Cohen, "'Her' and the Fairy Tale of Being Head-over-Heels with Tech," *Washington Post*, December 23, 2013, https://www.washingtonpost.com/opinions/richard-cohen-her-and-the-fairy-tale-of-being-head-over-heels-with-tech/2013/12/23/2a30a6ec-6bf8-11e3-a523-fe73f0ff6b8d_story.html, accessed September 13, 2020.

24. Emil Protalinski, "Amazon Echo Is a $200 Voice-Activated Wireless Speaker for Your Living Room," VentureBeat, November 6, 2014, https://venturebeat.com/2014/11/06/amazon-echo-is-a-200-voice-activated-smart-wireless-speaker-for-your-living-room/, accessed on March 21, 2019.

25. Hope King, "Amazon Built the Star Trek Computer for Your House," CNN Business, July 24, 2015, https://money.cnn.com/2015/07/24/technology/amazon-echo-review/index.html, accessed November 22, 2019.

26. "Amazon Echo Teardown Gets Inside the Smart Speaker Powered by the Cloud," CNET, September 14, 2015, https://www.cnet.com/news/amazon-echo-teardown-a-smart-speaker-powered-by-amazons-cloud/, accessed November 30, 2019.

27. Clive Thompson, "Watch What You Say: The Cloud Might Be Listening," *Wired*, November 20, 2015, https://www.wired.com/2015/11/clive-thompson-9/, accessed November 30, 2019.

28. Mike Troy, "The Meaning of Prime Day," *Retail Leader*, July 15, 2019, https://retailleader.com/meaning-prime-day, accessed September 2, 2019.

29. Nathan Olivarez-Giles, "Amazon Turns Its Alexa Virtual Assistant into a Prime Day Personal Shopper," *Wall Street Journal*, updated July 8, 2016, https://www.wsj.com/articles/amazon-turns-its-alexa-virtual-assistant-into-a-prime-day-personal-shopper-1468005812, accessed August 24, 2020.

30. Mike Troy, "The Meaning of Prime Day," *Retail Leader*, July 15, 2019, https://retailleader.com/meaning-prime-day, accessed September 2, 2019.

31. Day One Staff, "How and Why Prime Day Came to Be, and the Impact and Scale for Customers, Small Businesses, and Sellers on Amazon," Amazon Blog DayOne, https://blog.aboutamazon.com/shopping/the-history-of-prime-day, accessed September 1, 2019.

32. Monica Nickelsburg, "Could Smart Speakers, Like Amazon Echo, Outpace Early iPhone Sales?," GeekWire, November 2, 2016, https://www.geekwire.com/2016/age-alexa-amazon-echo-sales-pace-first-year-iphones/, accessed September 1, 2019.

33. Ibid.

34. Forbes Finds Contributor Group, "Amazon Prime 2018: Here's the Information You Need to Know," July 3, 2018, https://www.forbes.com/sites/forbes-finds/2018/07/03/prime-day/#2004b161622d, accessed November 30, 2019.

35. Ibid. According to Forbes, Forbes Finds "covers products we think you'll love. Featured products are independently selected and linked to for your convenience. If you buy something using a link on this page, Forbes may receive a small share of that sale."

36. Alivia McAtee, "A Prime Day Shopping List to Build out Your Smart Home," Reviews.com, July 12, 2019, https://www.reviews.com/blog/prime-day-smart-homes/, November 27, 2019.

37. Catey Hill, "Here Are Some of the Best-Selling Items on Amazon Prime Day from Around the World," MarketWatch, July 17, 2019, https://www.marketwatch.com/story/here-are-some-of-the-best-selling-items-on-amazon-prime-day-from-around-the-world-2019-07-17, accessed August 24, 2020.

38. Catey Hill, "Amazon Gave Away Thousands of These Devices Free with Purchase on Prime Day, and They're Still Doing It Today," MarketWatch, July 16, 2019, https://www.marketwatch.com/story/amazon-gave-away-thousands-of-these-devices-free-with-purchase-on-prime-day-and-theyre-still-doing-it-today-2019-07-16, accessed August 24, 2020.

39. "Big Board Today's Top Stories," ABC News Good Morning America, July 10, 2017.

40. Rick Broida, "Early Prime Day Deals on Google Hardware: $25 Home Mini, $79 Nest Hub and Lots More," CNETNews.com, July 10, 2019.

41. Simon Cohen, "Walmart's Got the Google Home Max for $249 Just in Time for Prime Day," Digital Trends, July 15, 2019, https://www.digitaltrends.com/home-theater/google-home-max-prime-day-2019-deal-walmart/, accessed September 2, 2019.

42. "Voicebot: More Than a Quarter of Americans Now Own Smart Speakers," RAIN News, March 7, 2019, https://rainnews.com/voicebot-more-than-a-quarter-of-americans-now-own-smart-speakers/, accessed September 3, 2019.

43. Bret Kinsella, "U.S. Smart Speaker Ownership Rises 40% in 2018 to 66.4 Million and Amazon Echo Maintains Market Share Lead Says New Report from Voicebot," Voicebot.ai, March 7, 2019, https://voicebot.ai/2019/03/07/u-s-smart-speaker-ownership-rises-40-in-2018-to-66-4-million-and-amazon-echo-maintains-market-share-lead-says-new-report-from-voicebot/, accessed September 3, 2019.

44. Dan Gallagher, "Sonos Can't Miss a Beat," *Wall Street Journal* Online, September 10, 2018, accessed via Factiva; Brian X. Cen and Daisuke Wakabayashi, "New Pixel Phones and Other Gadgets Keep Google in the Hardware Hunt," *New York Times*, October 9, 2018; Dan Gallagher, "Why Big Tech Keeps Trying Its Hand at Hardware," *Wall Street Journal* Online, October 19, 2018, via Factiva; Taylor Telford,

"Those News Anchors Are Professional and Efficient," *Washington Post*, November 10, 2018.

45. Clive Thompson, "May A.I. Help You?," *New York Times,* November 18, 2019, https://www.nytimes.com/interactive/2018/11/14/magazine/tech-design-ai-chatbot.html, November 25, 2019.

46. David Pierce, "Smart TVs Will Get Dumber," *Wall Street Journal*, November 26, 2018, p. B4.

47. Alexandra Petri, "A Ranking of 100—Yes, 100—Christmas Songs," *Washington Post,* https://www.washingtonpost.com/opinions/2018/12/07/ranking-yes-christmas-songs/, accessed November 30, 2019.

48. Caroline Knorr, "What Parents Need to Know Before Buying Google Home or Amazon Echo," *Washington Post*, December 14, 2018, https://www.washingtonpost.com/lifestyle/2018/12/14/what-parents-need-know-before-buying-google-home-or-amazon-echo/, accessed December 1, 2019.

49. Geoffrey Fowler, "I Live with Alexa, Google Assistant and Siri. Here's Which One You Should Pick," *Washington Post,* November 21, 2018, https://www.washingtonpost.com/technology/2018/11/21/i-live-with-alexa-google-assistant-siri-heres-which-you-should-pick/, accessed December 1, 2019; and Amy Webb, "Are You an Amazon or an Apple Family?," *New York Times,* March 9, 2019, https://www.nytimes.com/2019/03/09/opinion/sunday/10Webb.html, accessed December 1, 2019. Other media outlets ran similar stories. See, for example, "Which Voice Assistant Is Right for You?," February 12, 2018, "Good Morning America," February 12, 2018, https://www.goodmorningamerica.com/living/video/voice-assistant-53009925, accessed December 11, 2019; Wirecutter staff, "Smart-Home Devices to Make Your Holidays Easier," *New York Times,* October 26, 2018, https://www.nytimes.com/2018/10/26/smarter-living/wirecutter/smart-home-devices-holiday-tasks.html, accessed December 1, 2019; Wirecutter staff, "Gift Guide 2018 Smarter Home," *New York Times,* several dates with products constantly updated (despite the title, the link leads to a page with some 2019 smarter home products), https://www.nytimes.com/guides/gifts/2018-holiday-gift-guide?category=Smarter+Home&redirect=true, accessed December 11, 2019; Daniel

Bortz, "How Your Thermostat Could Save You a Bundle This Winter," *Washington Post,* November 28, 2018, https://www.washingtonpost. com/lifestyle/home/how-your-thermostat-could-save-you-a-bundle-this-winter/2018/11/27/a7d6cef2-ec4d-11e8-96d4-0d23f2aaad09_ story.html, accessed December 1, 2019; Laura Stevens, "Amazon Wants Alexa to Do More Than Just Play Your Music," *Wall Street Journal* Online, October 20, 2018, https://www.wsj.com/articles/ amazon-wants-alexa-to-do-more-than-just-play-your-music-1540047600, accessed August 24, 2020.

50. David Pierce, "33 Mostly Free Ways to Fix Your Family's Tech Problems," *Wall Street Journal* Online, December 23, 2018, https:// www.wsj.com/articles/33-mostly-free-ways-to-fix-your-familys-tech-problems-11545573601, accessed December 1, 2019.

51. James Burch, "In Japan, a Buddhist Funeral for Robot Dogs," *National Geographic,* May 24, 2018, https://www.nationalgeographic.com/ travel/destinations/asia/japan/in-japan--a-buddhist-funeral-service-for-robot-dogs/, accessed September 18, 2019.

52. Jaclyn Jeffret-Wilensky, "Why Robotic Pets May Be the Next Big Thing in Dementia Care," NBC News, April 3, 2019, https://www. nbcnews.com/mach/science/why-robotic-pets-dementia-care-may-be-next-big-thing-ncna990166, accessed September 18, 2019.

53. Judith Newman, "To Siri, with Love," *New York Times,* October 17, 2014, https://www.nytimes.com/2014/10/19/fashion/how-apples-siri-became-one-autistic-boys-bff.html, accessed December 1, 2019.

54. Mike Colias, "Ordering Coffee Through Your Car: New Apps Turn Cars into Smartphones, Raising Safety Questions," https://www.wsj. com/articles/ordering-coffee-through-your-car-new-apps-turn-cars-into-smartphones-raising-safety-questions-1537354800, accessed December 1, 2019.

55. Tony Romm, "Trump Administration Proposal Could Target Exports of the Tech Behind Siri, Self-Driving Cars, and Supercomputers," *Washington Post,* November 19, 2018, https://www.washingtonpost. com/technology/2018/11/19/trump-administration-proposal-could-target-exports-tech-behind-siri-self-driving-cars-supercomputers/, accessed December 1, 2019.

56. Ann-Marie Alcantara, "The Most Fun (and Useful) Things You Can Do with an Amazon Echo or Google Home," *New York Times,* September 19, 2018, https://www.nytimes.com/2018/09/19/smarter-living/fun-useful-things-amazon-echo-alexa-google-home.html, accessed December 1, 2019.

57. Thompson, "May A.I. Help You?"

58. Venessa Wong, "Amazon Knows Alexa Devices Are Laughing Spontaneously and It's 'Working to Fix It,'" BuzzFeed News, March 7, 2018, https://www.buzzfeednews.com/article/venessawong/amazon-alexa-devices-are-laughing-creepy, accessed December 1, 2019.

59. Thompson, "May A.I. Help You?"

60. Leah Fessler, "We Tested Bots Like Siri and Alexa to See Who Would Stand Up to Sexual Harassment," Quartz, February 27, 2017, https://qz.com/911681/we-tested-apples-siri-amazon-echos-alexa-microsofts-cortana-and-googles-google-home-to-see-which-personal-assistant-bots-stand-up-for-themselves-in-the-face-of-sexual-harassment/, accessed October 13, 2019.

61. Leah Fessler, "Amazon's Alexa Is Now a Feminist, and She's Sorry if That Upsets You," Quartz, January 17, 2018, https://qz.com/work/1180607/amazons-alexa-is-now-a-feminist-and-shes-sorry-if-that-upsets-you/, accessed October 13, 2019.

62. Ibid.

63. Madeline Buxton, "Writing for Alexa Becomes Complicated in the #MeToo Era," Refinery 20, December 27, 2019, https://www.refinery29.com/en-us/2017/12/184496/amazo-alexa-personality-me-too-era, accessed October 13, 2019.

64. Thao Phan, "Amazon Echo and the Aesthetics of Whiteness," *Catalyst* 5, no. 1 (2019): 23.

65. Drew Harwell, "Smart Speaker Revolution Raises the Issue of Accent Bias," *Washington Post,* July 20, 2018, p. A01.

66. Cade Metz, "There is a Racial Divide in Speech-Recognition Systems, Researchers Say," *New York Times,* March 23, 2020, https://www.nytimes.com/2020/03/23/technology/speech-recognition-bias-apple-amazon-google.html, accessed September 4, 2020.

67. Harwell, "Smart Speaker Revolution."

68. Ibid.

69. Niraj Chokshi, "Is Alexa Listening?," *New York Times,* May 25, 2018, https://www.nytimes.com/2018/05/25/business/amazon-alexa-conversation-shared-echo.html, accessed December 1, 2019.

70. Geoffrey Fowler, "There's a Spy in Your Home, and Its Name Is Alexa," *Washington Post*, May 12, 2019, G01.

71. Sapna Maheshwari, "Hey, Alexa, What Can You Hear? and What Will You Do with It?," *New York Times*, March 31, 2018, https://www.nytimes.com/2018/03/31/business/media/amazon-google-privacy-digital-assistants.html, December 1, 2019.

72. Ben Fox Rubin, "Amazon's New Alexa Features Put More Emphasis on Privacy," CNET, May 19, 2019, https://www.cnet.com/news/amazons-new-alexa-features-puts-added-emphasis-on-privacy/, accessed November 30, 2019.

73. Matt Day, Giles Turner, and Natalia Drozdiak, "Amazon Workers Are Listening to What You Tell Alexa," Bloomberg, April 10, 2019, https://www.bloomberg.com/news/articles/2019-04-10/is-anyone-listening-to-you-on-alexa-a-global-team-reviews-audio, accessed September 7, 2019.

74. Ibid.

75. "Strangers Are Listening to Everything You Ask Apple's Siri," CBS SF Bay Area, March 12, 2015, https://sanfrancisco.cbslocal.com/2015/03/12/strangers-apple-siri-data-third-party-privacy/, accessed September 7, 2019.

76. Alex Hern, "Apple Contractors 'Regularly Hear Confidential Details' on Siri Recordings," *Guardian*, July 26, 2019, https://www.theguardian.com/technology/2019/jul/26/apple-contractors-regularly-hear-confidential-details-on-siri-recordings, accessed September 7, 2019.

77. "Echo Dot Kids Edition Violates COPPA," Campaign for a Commercial-Free Childhood, May 9, 2019, https://www.echokidsprivacy.com/?eType=EmailBlastContent&eId=ca5d12fa-f830-4302-add5-3d73d87b84ab, accessed December 1, 2019.

78. Ben Fox Rubin, "Lawsuits Claim Amazon's Alexa Records Kids Without Their Consent," CNET, June 13, 2019, https://www.cnet.

com/news/lawsuits-claim-amazons-alexa-records-kids-without-their-consent/, accessed December 1, 2019.

79. Alfred Ng, "Amazon Alexa Transcripts Live On, Even after You Delete Voice Records," May 9, 2019, https://www.cnet.com/news/amazon-alexa-transcripts-live-on-even-after-you-delete-voice-records/, accessed December 1, 2019.

80. "Apple Statement: Improving Siri's Privacy Protections," Apple Newsroom, August 28, 2019, https://www.apple.com/newsroom/2019/08/improving-siris-privacy-protections/, accessed September 7, 2019.

81. Ibid.

82. Nick Bastone, "Google Will Temporarily Stop Contractors from Listening to Assistant Recordings Around the World after Leaked Data Sparked Privacy Concerns," *Business Insider*, August 2, 2019, https://www.businessinsider.com/google-stops-audio-reviews-europe-privacy-concerns-2019-8, accessed September 8, 2019.

83. Ibid.

84. Mary Hanbury, "Google Says Its Workers Are Listening to and Transcribing Your Google Assistant Commands," *Business Insider*, July 11, 2019, https://www.businessinsider.com/google-workers-listen-to-google-assistant-commands-2019-7, accessed September 8, 2019.

85. See Brian Koerber, "Amazon Reveals Why Alexa Is Randomly Laughing and Creeping Out People," Mashable, March 7, 2018, https://mashable.com/2018/03/07/why-amazon-alexa-laughing/; Wong, "Amazon Knows Alexa Devices Are Laughing Spontaneously"; and "Amazon Working to Fix Alexa's Laughing Problem," March 8, 2018, https://www.goodmorningamerica.com/news/video/amazon-working-fix-alexas-laughing-problem-53604568, accessed December 11, 2019.

86. Hayley Tsukayama, "How Closely Is Amazon's Echo Listening?," *Washington Post*, November 11, 2014, https://www.washingtonpost.com/news/the-switch/wp/2014/11/11/how-closely-is-amazons-echo-listening/, accessed June 1, 2010.

87. Natasha Singer, "Just Don't Call It Privacy," *New York Times*, September 22, 2018, https://www.nytimes.com/2018/09/22/sunday-review/privacy-hearing-amazon-google.html, accessed June 1, 2020.

88. See "Amazon Privacy Notice," Amazon, August 29, 2017, https://
www.amazon.com/gp/help/customer/display.
html?ie=UTF8&nodeId=16015091; and Alexa and Alexa Device FAQ,"
n.d., https://www.amazon.com/gp/help/customer/display.
html?nodeId=201602230, accessed August 16, 2019. The updated
privacy notice of January 2020 does indirectly assert Amazon's right
to advertise to its users based on what they say to Alexa (see p. 126).
The company does not present this activity straightforwardly even in
its FAQs. As of July 2020, the word "advertising" appears there only—
to assure readers that Amazon does "not sell children's personal information for advertising or other purposes."

89. Chris Burt, "Amazon Hit by Illinois Biometric Data Privacy Suit for
Alexa Recordings," July 8, 2019, BiometricUpdate.com, https://www.
biometricupdate.com/201907/amazon-hit-by-illinois-biometric-data-
privacy-suit-for-alexa-recordings, accessed December 7, 2019.

90. Emily Birnbaum, "Advocacy Groups Press Congress to Probe
Amazon's 'Surveillance Empire,'" *The Hill*, November 25, 2019,
https://thehill.com/policy/technology/471903-civil-rights-groups-
press-for-congressional-investigation-into-amazons, accessed
November 30, 2019.

91. Caroline Haskins, "Amazon's Home Security Company Is Turning
Everybody into Cops," *Vice*, February 7, 2019, https://www.vice.com/
en_us/article/qvyvzd/amazons-home-security-company-is-turning-
everyone-into-cops, accessed November 30, 2019.

92. Birnbaum, "Advocacy Groups Press Congress."

93. Emily Birnbaum, "Activists Form National Coalition to Take on
Amazon," *The Hill*, November 26, 2019, https://thehill.com/policy/
technology/472153-activists-form-national-coalition-to-take-on-
amazon, accessed December 1, 2019.

94. Fowler, "There's a Spy in Your Home."

95. Ibid.

96. For example, Bastone, "Google Will Temporarily Stop Contractors";
Hanbury, "Google Says Its Workers Are Listening"; see also Koerber,
"Amazon Reveals Why Alexa Is Randomly Laughing"; Wong,
"Amazon Knows Alexa Devices Are Laughing Spontaneously";

"Amazon Working to Fix Alexa's Laughing Problem,"; and Tsukayama, "How Closely Is Amazon's Echo Listening?"

97. Sonia Rao, "In Today's Homes, Consumers Are Willing to Sacrifice Privacy for Convenience," *Washington Post* Online, September 12, 2018, via Factiva.

98. "Smart Speaker Consumer Adoption Report," Voicebot.ai and Voicify, March 2019, https://voicebot.ai/wp-content/uploads/2019/03/smart_speaker_consumer_adoption_report_2019.pdf, accessed December 1, 2019.

99. Meenakshi Tiwari, "Voice Technology Is the Key to Faster Smart Home Adoption," Forrester, May 7, 2019, https://go.forrester.com/blogs/voice-technology-is-the-key-to-faster-smart-home-adoption/, accessed August 24, 2020.

5 ADVERTISERS GET READY

1. Alexei Kounine, "5 Things You Need to Know About the Voice-Activated Future of Marketing," Marketing Insider, February 20, 2019, https://www.mediapost.com/publications/article/332219/5-things-you-need-to-know-about-the-voice-activate.html, accessed June 19, 2019.

2. This information came through various interviews, trade magazine reading, and informal discussions with executives and voice-software developers at the Voice 2010 trade conference.

3. Trefis Team, "Is Google Advertising Revenue 70%, 80%, or 90% of Alphabet's Total Revenue?," *Forbes,* December 24, 2019, https://www.forbes.com/sites/greatspeculations/2019/12/24/is-google-advertising-revenue-70-80-or-90-of-alphabets-total-revenue/#4e245614a01c, accessed April 14, 2010.

4. Ginny Marvin, "Amazon Ad Revenue Tops $3.5 Billion in Third Quarter, Expecting Strong Holiday Season," Marketing Land, https://marketingland.com/amazon-ad-revenue-tops-3-5-billion-in-third-quarter-expecting-strong-holiday-season-269735, accessed April 14, 2020.

5. Three developers at the Voice 2019 conference confirmed that they receive transcripts of what people say in relation to their voice apps.

6. Kounine, "5 Things You Need to Know."

7. Glen Shires et al. for Google LLC, "Speech Recognition and Summarization," US Patent 10,185,711 B1, January 22, 2019, column 8 (at line 38), https://patents.google.com/patent/US8612211, accessed August 24, 2020.

8. Ibid.

9. Collin et al. for Amazon Technologies, Inc., "Cross-Channel Online Advertising Attribution," US Patent 10,169,778 B1, January 1, 2019 column 2 (in abstract), https://patents.google.com/patent/US10169778B1/en, accessed August 24, 2020.

10. Max Willens, "How 3 Publishers Are Staffing for Amazon Echo," Digiday, October 12, 2016, https://digiday.com/careers/three-publishers-staffing-amazon-echo/, accessed June 19, 2019.

11. Matt Weinberger, "How Amazon's Echo Went from a Smart Speaker to the Center of Your Home," Business Insider, May 23, 2017, https://www.businessinsider.com/amazon-echo-and-alexa-history-from-speaker-to-smart-home-hub-2017-5, accessed June 19, 2019.

12. VoiceLabs Gives Brands the Ability to Reach and Interact with Amazon Echo Consumers, While Protecting the Consumer Experience," PR Newswire, May 11, 2017, https://www.prnewswire.com/news-releases/voicelabs-gives-brands-the-ability-to-reach-and-interact-with-amazon-echo-consumers-while-protecting-the-consumer-experience-300455963.html, accessed June 19, 2019.

13. Email communication from Nick Schwab, a computer engineer who created Amazon skills and intended to use VoiceLabs to make money from advertising, June 23, 2010.

14. Bret Kinsella, "Amazon Adds Further Restrictions to Alexa Skill Advertising," Voicebot.ai, April 20, 2017, https://voicebot.ai/2017/04/20/amazon-adds-restrictions-alexa-skill-advertising/, accessed June 25, 2019.

15. R. Y. Christ, "VoiceLabs Is Putting Interactive Ads into Your Alexa Skills," CNET, May 11, 2017, https://www.cnet.com/news/voicelabs-is-putting-interactive-ads-into-your-alexa-skills/, accessed June 19, 2019.

16. Sapna Maheshwari, "Burger King 'O.K. Google' Ad Doesn't Seem OK with Google," New York Times, April 12, 2017, https://www.nytimes.

com/2017/04/12/business/burger-king-tv-ad-google-home.html, accessed June 20, 2019.

17. Sarah Perez, "Amazon's New Alexa Developer Policy Bans All Ads Except in Music and Flash Briefings," TechCrunch, April 20, 2017, https://techcrunch.com/2017/04/20/amazons-new-alexa-developer-policy-now-bans-all-ads-except-in-music-and-flash-briefings/, accessed June 19, 2019.

18. Khari Johnson, "VoiceLabs Suspends Its Amazon Alexa Skill Ad Network," VentureBeat, June 16, 2017, https://venturebeat.com/2017/06/15/voicelabs-suspends-amazon-alexa-skill-ad-network/, accessed June 20, 2019.

19. Bret Kinsella, "Amazon Alexa Skill Counts Rise Rapidly in the U.S., U.K., Germany, France, Japan, Canada, and Australia," Voicebot.ai, January 2, 2019, https://voicebot.ai/2019/01/02/amazon-alexa-skill-counts-rise-rapidly-in-the-u-s-u-k-germany-france-japan-canada-and-australia/, accessed June 24, 2019.

20. Bret Kinsella, "Google Assistant Actions Total 4,253 in January 2019, Up 2.5x in Past Year but 7.5% the Total Number Alexa Skills in U.S," Voicebot.ai, February 15, 2019, https://voicebot.ai/2019/02/15/google-assistant-actions-total-4253-in-january-2019-up-2-5x-in-past-year-but-7-5-the-total-number-alexa-skills-in-u-s/, accessed June 24, 2019.

21. Quotations from David Isbitski in this and the following paragraphs are from my interview with him on June 18, 2019.

22. Chris Davies, "Now Alexa Skills Can Sell You Stuff," Slash Gear, May 3, 2018, https://www.slashgear.com/amazon-alexa-in-skill-purchasing-amazon-pay-voice-integration-03529490/, accessed November 1, 2019.

23. Chris Davies, "Alexa Donations Turns Echo into Charity Collections Box," Slash Gear, April 2, 2018, https://www.slashgear.com/alexa-donations-turns-echo-into-charity-collection-box-02525580/, accessed November 1, 2019.

24. This information came through various interviews, notably with David Isbitski and Michael Dobbs; trade magazine reading; and informal discussions with executives and voice-software developers at the Voice 2019 trade conference.

25. Jennifer Wise, "Digital Voice Experiences," Forrester, March 27, 2019 (updated April 2, 2019), p. 2.

26. Ibid. pp. 1–2.

27. Interview with Joe Maceda, June 11, 2019.

28. Ibid.

29. Interview with Michael Dobbs, June 6, 2019.

30. Interview with Kirk Drummond, June 12, 2019.

31. Sarah Perez, "Voice Shopping Estimated to hit $40+ Billion Across U.S. and U.K. by 2022," TechCrunch, March 2, 2018, https://techcrunch.com/2018/03/02/voice-shopping-estimated-to-hit-40-billion-across-u-s-and-u-k-by-2022/, accessed June 28, 2019; also, OC&C Strategy Consultants, "Voice Set to Jump to $40 Billion by 2022, Rising from $2 Billion Today," *Business Insider,* PR Newswire press release, https://markets.businessinsider.com/news/stocks/voice-shopping-set-to-jump-to-40-billion-by-2022-rising-from-2-billion-today-1017434300, accessed August 24, 2020.

32. Greg Sterling, "Report: Amazon Internal Data Suggest 'Voice Commerce' Virtually Nonexistent," Marketing Land, August 8, 2018, https://marketingland.com/report-amazon-internal-data-suggest-voice-commerce-virtually-nonexistent-245664#, accessed June 26, 2019.

33. Ross Benes, "Few People Regularly Make Purchases Through Smart Speakers," eMarketer, December 20, 2018, https://content-na1.emarketer.com/few-people-regularly-make-purchases-through-voice-assistants, accessed June 27, 2019.

34. Greg Sterling, "As CES Opens, Amazon, Google Tout Digital Assistant Stats That Highlight Respective Market Strengths," Marketing Land, January 7, 2019, https://marketingland.com/as-ces-opens-amazon-google-tout-digital-assistant-stats-that-highlight-their-respective-market-strengths-254849, accessed June 26, 2019.

35. Perez, "Voice Shopping"; OC&C Strategy Consultants, "Voice Set to Jump to $40 Billion by 2022."

36. Isbitski interview.

37. Quoted in Tess Townsend, "Google Assistant Will Make Money from e-Commerce," Vox Recode, May 23, 2017, https://www.vox.com

/2017/5/23/15681596/google-assistant-ecommerce-revenue, accessed June 26, 2019.

38. Ibid.

39. Nat Ives, "Pandora Pitches Ads Targeted to Amazon and Google Smart Speakers," *Wall Street Journal*, March 13, 2019, https://www.wsj.com/articles/pandora-pitches-ads-targeted-to-amazon-and-google-smart-speakers-11552471201, accessed December 14, 2019.

40. Katie Nichol, "Interview with L'Oréal Chief Digital Officer Lubomira Rochet," BW Confidential, May 14, 2019, https://www.bwconfidential.com/interview-with-loreal-chief-digital-officer-lubomira-rochet/, accessed August 24, 2020.

41. Ad Age Staff, "Ad Age's 2019 Industry Predictions," *Ad Age*, January 8, 2019, https://adage.com/article/cmo-strategy/ad-age-s-2019-industry-predictions/316142, accessed September 7, 2020.

42. Maceda interview.

43. Local Search Engine Optimisation," Wikipedia n.d., https://en.wikipedia.org/wiki/Local_search_engine_optimisation, accessed June 28, 2019.

44. "Search Quality Evaluator Guidelines," Google, May 16, 2019, https://static.googleusercontent.com/media/www.google.com/en//insidesearch/howsearchworks/assets/searchqualityevaluatorguidelines.pdf; and "Bing Keyword Search," Bing, n.d., https://www.bing.com/toolbox/keywords, accessed June 30, 2019.

45. Chaitanya Chandrasekar, "Voice Search Optimization: 6 Big Changes You'll Need to Make," SEJ: Search Engine Journal, July 11, 2018, https://www.searchenginejournal.com/voice-search-optimization-changes/259975/#close, accessed June 30, 2019.

46. Interview with Pete Erickson, June 6, 2019.

47. Bradley Shaw, "Voice Search Statistics, Facts, and Trends 2019 for Online Marketers," SEOExpert, n.d., https://seoexpertbrad.com/voice-search-statistics/, accessed June 30, 2019.

48. Bradley Shaw, "How to Optimize for Voice Search," SEOExpert, n.d., https://seoexpertbrad.com/voice-search-optimization/, accessed June 30, 2019.

49. Maceda interview.

50. Tony Landa, "How We Are Training Alexa to Think for Herself," *Ad Age*, April 9, 2020, https://adage.com/article/digital/training-alexa/313039, accessed September 10, 2020.

51. Dobbs interview.

52. Ibid.

53. Interview with Rishad Tobaccowala, July 3, 2019. The other five executives are Michael Dobbs, Austin Arensberg, Kirk Drummond, Joe Maceda, and Janet Levine.

54. Interview with Will Margaritis, June 14, 2019.

55. Interview with Austin Arensberg, June 4, 2019.

56. Interview with Kirk Drummond, June 12, 2019.

57. Maceda interview.

58. Ibid.

59. Interview with Janet Levine, July 3, 2019.

60. Ibid.

61. Interview with Arafel Buzan, July 3, 2019.

62. David Mullen, "Mattersight Issued Four New Patents Signaling the Future of Analytics," Mattersight, January 9, 2018, https://www.globenewswire.com/news-release/2018/01/09/1286200/0/en/Mattersight-Issued-Four-New-Patents-Signaling-the-Future-of-Analytics.html; see also David Gustafson and Christopher Danson (for the Mattersight Corporation), "Personality-Based Chatbots and Methods," US Patent 9,847,084, December 19, 2017, https://patents.google.com/patent/US9847084B2/en, accessed August 24, 2020.

63. Khari Johnson, "Mattersight Wants to Use AI and Alexa to Send You Ads Based on Your Personality," VentureBeat, March 31, 2017, https://venturebeat.com/2017/03/31/mattersight-wants-to-use-ai-and-alexa-to-send-you-ads-based-on-your-personality/, accessed July 4, 2019.

64. Ibid.

65. Ibid.

66. Megan Graham, "Hey, Alexa: What's the Best Voice Strategy for Brands?," *Advertising Age*, October 18, 2017, https://adage.com/article/digital/hey-alexa-voice-strategy-brands/310893, accessed July 2, 2019; Maceda interview.

67. Maceda interview.

68. Interview with Mykolas Rambus, June 17, 2019.

69. Interview with Brandon Purcell, June 21, 2019.

70. Dobbs interview.

71. Interview with Joe Petro, July 19, 2019.

72. Interview with Ethan Goodman, August 21, 2019.

73. P. V. Kannan and Josh Bernoff, "Here's Why You're Not Using Facebook Messenger or Amazon Alexa to Call Customer Support," MarketWatch, June 21, 2019, https://www.marketwatch.com/story/heres-why-youre-not-using-facebook-messenger-or-amazons-alexa-to-call-customer-support-2019-06-21, accessed December 1, 2019.

74. Ibid.; interview with Bret Kinsella, April 19, 2019.

75. Bret Kinsella, "Voice Insider #28: Why Alexa and Google Will Spawn Thousands of Assistants," *Voice Insider*, March 1, 2019, https://www.patreon.com/posts/25062829, accessed September 19, 2019.

76. Ibid.; see also Bret Kinsella, "Voice Insider #25: "The GOWN Voice Assistant Classification System Framework Debut," *Voice Insider*, February 7, 2019, https://www.patreon.com/posts/voice-insider-25-24540972, accessed September 19, 2019.

77. Quoted in Bret Kinsella, "Pandora Taps Instreamatic to Test Voice-Enabled Ads," Voicebot.ai, April 3, 2019, https://voicebot.ai/2019/04/03/pandora-taps-intreamatic-to-test-voice-enabled-ads/, accessed December 14, 2019.

78. Eric Hal Schwartz, "Pandora Begins Running Interactive Voice Ads," Voicebot.ai, https://voicebot.ai/2019/12/13/pandora-begins-running-interactive-voice-ads/, accessed December 14, 2019.

79. Brian Roemmele, "Meet Erica, Bank of America's New Voice AI System," *Forbes*, October 28, 2016, https://www.forbes.com/sites/quora/2016/10/28/meet-erica-bank-of-americas-new-voice-ai-banking-system/#3cc93df350db, accessed September 19, 2019.

80. "Erica," Bank of America, n.d., https://promo.bankofamerica.com/erica/, accessed September 19, 2019.

81. Penny Crossman, "Mad About Erica: Why a Million People Use Bank of America's Chatbot," American Banker, https://www.american-

banker.com/news/mad-about-erica-why-a-million-people-use-bank-of-americas-chatbot, accessed September 19, 2019.

82. "Bank of America's Erica Completes More Than 50 Million Client Requests in First Year," Bank of America press release via Business Wire, https://www.businesswire.com/news/home/20190528005646/en/Bank-America%E2%80%99s-Erica%C2%AE-Completes-50-Million-Client, accessed September 19, 2019.

83. Pete Erickson replying to @BretKinsella on Twitter, August 27, 2019, https://twitter.com/bretkinsella/status/1166318384021159937, accessed September 19, 2019.

84. Interview with Pete Erickson, June 12, 2019.

85. Purcell interview.

86. See, for example, Brittany Page, "Hey Alexa, How Do I Get My Product Visible in Amazon Search in 2019?," Search Engine Land, January 2, 2019, https://searchengineland.com/hey-alexa-how-do-i-get-my-product-visible-in-amazon-search-in-2019-309864, accessed July 1, 2019.

6 VOICE PROFILING AND FREEDOM

1. Thanks to William Frucht for help in formulating these examples.

2. "What About Advertising?," Amazon Privacy Notice, January 1, 2020, https://www.amazon.com/gp/help/customer/display.html/ref=hp_bc_nav?ie=UTF8&nodeId=201909010#GUID-1B2BDAD4–7ACF-4D7A-8608-CBA6EA897FD3__SECTION_87C837F9CCD84769B4AE2BEB14AF4F01, accessed November 21, 2020.

3. "Our Commitment to Privacy in the Home," Google Nest Privacy, https://store.google.com/category/google_nest_privacy, accessed September 19, 2019, and July 16, 2020.

4. Google Privacy Policy, July 1, 2020, https://policies.google.com/privacy/embedded?hl=en-US, accessed July 16, 2020.

5. "The Mars Agency Introduces Marilyn[SM], the First End-to-End, AI-Enabled Predictive Commerce Intelligence Platform for Marketing to Shopper," Mars Agency, June 11, 2019, https://www.themarsagency.com/news-article?title=new-article-introducing-marilyn, accessed April 16, 2020.

6. Jackie Snow, "In the AI Era, Your Voice Could Give Away Your Face," *Fast Company*, June 6, 2019, https://www.fastcompany. com/90357561/this-ai-guesses-human-faces-based-only-on-their-voices, accessed September 25, 2019.

7. See Jane Martin, "What Should We Do with a Hidden Curriculum When We Find One?," *Curriculum Inquiry* 6, no. 2 (1976): 135–151.

8. Quoted in ibid.

9. George Gerbner, "The Teacher Image and the Hidden Curriculum," *American Scholar* 42 (1973): 71.

10. Michael McGerr, *The Decline of Popular Politics: The American North, 1865–1928* (New York: Oxford University Press, 1986), p. 94.

11. Much of this material is taken from Joseph Turow, Michael X. Delli Carpini, Nora Draper, and Rowan Howard Williams, "Americans Roundly Reject Tailored Political Advertising," Annenberg School for Communication Departmental Papers, July 2012, pp. 5–6.

12. "Access to and Use of Voter Registration Lists," National Conference of State Legislatures, August 5, 2019, http://www.ncsl.org/research /elections-and-campaigns/access-to-and-use-of-voter-registration-lists.aspx, accessed September 23, 2019.

13. Jack Corrigan, "DHS Is Collecting Biometrics on Thousands of Refugees Who Will Never Enter the U.S.," Nextgov, August 20, 2019, https://www.nextgov.com/emerging-tech/2019/08/dhs-collecting-biometrics-thousands-refugees-who-will-never-enter-us/159310/, accessed September 24, 2019.

14. Jonathan Cantor and Donald K. Hawkins, "Privacy Impact Assessment for the United Nations High Commissioner for Refugees (UNHCR) Information Data Share," US Department of Homeland Security, DHS USCIS/PIA-081, August 13, 2019, p. 4, https://www.dhs.gov/sites/ default/files/publications/privacy-pia-uscis081-unhcr-august2019.pdf, accessed September 24, 2019.

15. For more on this topic, see Btihaj Ajana, *Governing Through Biometrics: The Biopolitics of Identity* (New York: Palgrave MacMillan, 2013), especially pp. 62–77.

16. George Joseph and Debbie Nathan, "Prisons Across the U.S. Are Quietly Building Databases of Incarcerated of Incarcerated People's

Voice Prints," *Appeal,* January 30, 2019, https://theappeal.org/prisons-across-the-u-s-are-quietly-building-databases-of-incarcerated-peoples-voice-prints/, accessed September 26, 2019.

17. Ibid.

18. "The California Consumer Privacy Act of 2018" (Assembly Bill No. 375), par. 1798.140b, June 28, 2018, https://leginfo.legislature.ca.gov/faces/billTextClient.xhtml?bill_id=201720180AB375, accessed September 24, 2019.

19. "Pandora Privacy Policy," Pandora, January 1, 2020, https://www.pandora.com/privacy, accessed January 1, 2020.

20. "Find Your Voice Mode," Pandora, n.d., https://www.pandora.com/voicemode/#faqs, accessed July 17, 2020.

21. Thanks to Professor Chris Jay Hoofnagle of UC Berkeley Law School for this important point.

22. "Explore Our FAQs," Bank of America, n.d., https://promo.bankofamerica.com/erica/, accessed September 7, 2020.

23. Julia Stead, "Conversations First: How Financial Services Is Evolving for the Voice Era," *Payments Journal*, February 2, 2018, https://www.paymentsjournal.com/conversations-first-financial-services-evolving-voice-era/, accessed July 17, 2020.

24. "Bank of America US Online Privacy Notice," https://www.bankofamerica.com/security-center/privacy-overview/#privacyNoticeBanner, accessed September 7, 2020.

25. Illinois General Assembly, "(740 ILCS 14/) Biometric Information Privacy Act," effective October 3, 2008, http://www.ilga.gov/legislation/ilcs/ilcs3.asp?ActID=3004&ChapterID=57, accessed December 7, 2019.

26. Robert Fallah, "Illinois Supreme Court Ruling: Biometric Privacy Law Only Requires Violation, Not Actual Harm," Fisher Phillips, https://www.fisherphillips.com/Employment-Privacy-Blog/illinois-supreme-court-ruling-biometric-privacy-law#:~:targetText=Violations%20of%20BIPA%20incur%20a,or%20recklessly%20violating%20the%20Act., accessed December 8, 2019.

27. Joseph Turow, Michael Hennessy, and Nora Draper, "Persistent Misperceptions: Americans' Misplaced Confidence in Privacy Policies,

2003–2015," *Journal of Broadcasting & Electronic Media* 62, no. 3 (2018): 461–478. The 2019 survey was carried out by Pew Research Center.

28. See Joseph Turow, *Breaking Up America: Advertisers and the New Media World* (Chicago: University of Chicago Press, 1997).

29. Bret Kinsella, "Baidu Is Reshaping Smart Speaker and Voice Assistant Market Share in China," Voicebot.ai, September 5, 2019, https://voicebot.ai/2019/09/05/baidu-is-reshaping-smart-speaker-and-voice-assistant-market-share-in-china/, accessed September 27, 2019.

30. Eric Hall Schwartz, "Baidu Upgrades DuerOS Voice Platform and Hits 400M Device Milestone," Voicebot.ai, July 3, 2019, https://voicebot.ai/2019/07/03/baidu-upgrades-dueros-voice-platform-and-hits-400m-device-milestone/, accessed September 27, 2019.

31. Kinsella, "Baidu."

32. Samm Sacks and Lorand Laskai, "China's Privacy Conundrum," *Slate*, February 7, 2019, https://slate.com/technology/2019/02/china-consumer-data-protection-privacy-surveillance.html, accessed September 27, 2019.

33. "Baidu Chief under Fire for Privacy Comments," March 28, 2018, *People's Daily Online* [English], http://en.people.cn/n3/2018/0328/c90000-9442509.html, September 27, 2019.

34. Sacks and Laskai, "China's Privacy Conundrum."

35. Ibid.

36. "Four Ministries Will Rectify App Personal Information to Collect Chaotic Violations or Will Revoke Business Licenses," Tencent [translated from Chinese to English by Google Translate], January 25, 2019, https://new.qq.com/omn/20190125/20190125A0CZW7.html, accessed September 27, 2019.

37. Sacks and Laskai, "China's Privacy Conundrum."

38. Ibid.

39. See "General Data Protection Regulation," Regulation (EU) 2016/679 of the European Parliament and of The Council, April 27, 2016, preamble, https://eur-lex.europa.eu/eli/reg/2016/679/oj, accessed August 24, 2020.

40. Jennifer Baker, "How the ePrivacy Regulation Talks Failed . . . Again," IAPP.org [website for International Association of Privacy

Professionals], November 26, 2019, https://iapp.org/news/a/how-the-eprivacy-regulation-failed-again/, accessed June 7, 2020.

41. "ePrivacy Regulation (European Union)," Wikipedia, https://en.wikipedia.org/wiki/EPrivacy_Regulation_(European_Union)#, accessed December 8, 2019.

42. "General Data Protection Regulation," article 4: definitions, item 14.

43. Ibid., article 9, paragraph 1.

44. Ibid., article 9, paragraph 2a.

45. Mania Aslan, "How Do the Rules on Audio Recording Change Under the GDPR?," IAPP.com [website for International Association of Privacy Professionals], April 24, 2018, https://iapp.org/news/a/how-do-the-rules-on-audio-recording-change-under-the-gdpr/, accessed December 8, 2019.

46. Quoted in Lindsey O'Donnell, "Amazon Alexa, Google Home on Collision Course with Regulation," Threatpost, https://threatpost.com/amazon-alexa-google-home-regulation/146357/, accessed December 8, 2019.

47. Ibid.

48. Marcus Hoy, "Telecom Voice Recording Ban Is Test Case for EU Privacy Rules," *Bloomberg Law,* April 6, 2019, https://news.bloomberglaw.com/privacy-and-data-security/voice-recording-opt-out-mandate-seen-as-privacy-wake-up-call, December 8, 2019.

49. "General Data Protection Regulation," article 9, paragraph 2a.

50. Chris Jay Hoofnagle, Bart van der Sloot, and Frederik Zuiderveen Borgesius, "The European Union General Data Protection Regulation: What It Is and What It Means," *Information & Communication Technology Law*, February 2019, https://www.tandfonline.com/doi/full/10.1080/13600834.2019.1573501, accessed December 13, 2019.

51. Lauren Bass, "The Concealed Cost of Convenience: Protecting Personal Data: Privacy in the Age of Alexa," *Fordham Intellectual Property, Media and Entertainment Law Journal* 30, no. 261 (2019): 314–318.

52. Ibid., pp. 264–265.

53. "FTC Fact Sheet: It's the Law," n.d., Federal Trade Commission, https://www.consumer.ftc.gov/sites/default/files/games/off-site/

youarehere/pages/pdf/FTC-Ad-Marketing_The-Law.pdf, accessed
April 19, 2020.

54. R. B. Mikkelsen, M. Gjerris, and G. Waldemar, "Broad Consent for
Biobanks Is Best—Provided It Is Also Deep," *BMC Medical Ethics* 20,
no. 71 (2019), https://doi.org/10.1186/s12910-019-0414-6, accessed
April 17, 2020.

55. Quoted in Dave Gershgorn, "Here's How Amazon Alexa Will
Recognize When You're Frustrated," OneZero, September 27,
2019, https://onezero.medium.com/heres-how-amazon-alexa-
will-recognize-when-you-re-frustrated-a9e31751daf7, accessed
September 29, 2019.

56. Quoted in ibid.

INDEX

Hicks, Mar, 267
hidden curriculum, 236–237
Hill, Jonah, 158
Home (smart speaker), 97, 166, 173, 229; advertising restricted on, 194, 196, 209; launch of, 3, 56; marketing of, 60, 63–64; market share of, 167; in neighborhood surveillance networks, 105–106; pricing of, 115, 166, 251; privacy policies of, 250; wide availability of, 2
home construction, 122
Home Depot, 116
Homeland Security Act (2002), 244–245
Home Max (smart speaker), 167
HomePod (smart speaker), 56–57, 60, 166, 167, 181
Honeywell International, 121, 122
Hoofnagle, Chris, 259
hormones, 79, 80
Horvitz, Eric, 155
hotels, 138–142

IBM Corporation, 23, 67–68, 221
IHeartRadio, 194
IKEA, 224
Illinois Biometric Information Privacy Act (BIPA), 185, 248–250
implicit discovery, 211
industrial construction of audiences, 24
Industrial Revolution, 14

InfoCision, 51
Instagram, 19, 89
insurance rates, 133
interactive voice distribution (IVR), 44–45, 67, 74
internet-driven private branch exchange (iPBX), 49
Invoca (voice analytics firm), 86, 248
iOS (operating system), 57
iPhone, 54, 57, 64, 135, 164
Isbitski, David, 200–201, 206, 207

Jackson, Philip, 236
Jackson, Samuel L., 62
Jetson Agency, 149
Jobs, Steve, 54
Johansson, Scarlett, 158
Johnson, Khari, 216
Jonze, Spike, 158
Juniper Consulting Group, 21

Kahler, Taibi, 94
Katz, James, 31
KB Homes, 121, 122
Kennedy, John F., 238
Kim, Ed, 128
Kindle, 55, 162
King, Hope, 160
Kinsella, Bret, 149–150, 197, 224–225; China's future viewed by, 254; compatible devices viewed by, 129; personalization viewed by, 30; smart speaker growth noted

Ritchie, Paul, 42
Roberts, Sarah, 31
robotic pets, 170–171
Rochet, Lubomira, 207
Roddenberry, Gene, 59
Roku (digital player manufacturer), 121
Romney, Mitt, 238
Ruckus Security, 122
Russell, Brad, 115–116, 120

Sacks, Samm, 254–256
Samsung, 27, 121, 123, 236; Bixby, 3, 9, 57, 62, 119; Galaxy, 156
Saturday Evening Post, 14
schizophrenia, 80
Schlage, 122
Schmidt, Eric, 20
schools, voice technology in, 142–145
Scrum Ventures, 213
search engine optimization (SEO), 209–210
Securus Technologies, 242
seductive surveillance, 11–12, 28–31, 35, 37, 63, 69, 251
segmentation, 16–17
Sejnoha, Vlad, 42–43
self-driving cars, 129
sentiment, of customers, 1, 7, 24, 90–93, 95, 228, 231, 267
sexual harassment, 173–174
Shaw, Bradley, 210
Shea Homes, 121, 123–124

shopping, 13–14, 23, 192; by Amazon Prime members, 34, 35, 56, 57–58, 113–115, 159, 162–167, 188; consumers' attitudes toward, 32, 229; with mobile devices, 55; personalization in, 45–50; predictive analytics and, 20; with virtual assistants, 147–150, 205, 220, 234
Singh, Rita, 78–79, 80–83
Siri (virtual assistant), 2, 3, 9, 34, 41, 56, 97, 116; Alexa vs., 156–157; autistic children and, 171–172; automotive applications of, 128; criticisms of, 62, 156–157, 180; fictional analogues of, 157–159; humanization of, 59–60; inappropriate responses from, 173; launch of, 153, 227; media's hyperventilating over, 153–156; origins of, 53, 54; public acceptance of, 53; search algorithms of, 211; as shopping assistant, 145
SmartAisle, 147–148
smartphones, 2, 97, 145, 153
smart speakers, 2, 3, 23, 97, 98, 138, 144; displays integrated into, 218, 235; market penetration of, 167, 203, 208; pervasiveness of, 169; purchases through, 205; race and, 174–176, 228–229;